IET TRANSPORTATION SERIES 17

Smart Sensing for Traffic Monitoring

Other volumes in this series:

Smart Sensing for Traffic Monitoring

Edited by
Nobuyuki Ozaki

The Institution of Engineering and Technology

Published by The Institution of Engineering and Technology, London, United Kingdom

The Institution of Engineering and Technology is registered as a Charity in England & Wales (no. 211014) and Scotland (no. SC038698).

© The Institution of Engineering and Technology 2021

First published 2020

The Institution of Engineering and Technology
Michael Faraday House
Six Hills Way, Stevenage
Herts, SG1 2AY, United Kingdom

www.theiet.org

British Library Cataloguing in Publication Data
A catalogue record for this product is available from the British Library

ISBN 978-1-78561-774-4 (hardback)
ISBN 978-1-78561-775-1 (PDF)

Typeset in India by MPS Limited

Contents

About the editor

Nobuyuki Ozaki is currently a professor at Nagoya University. His research interests focus on intelligent transport systems, or ITS, including new sensing approaches deploying image recognition technologies. He had worked for Toshiba Corporation more than 30 years and his last position was a senior fellow for promoting new technologies into ITS. He is also actively working for several ITS industry organizations both in Japan and internationally. He is serving as the vice president of Standard and Industrial Relations in the IEEE ITS Society.

Preface

Urbanization has become a worldwide trend, and it is anticipated to reach 60% of the world in the early 2030s. The percentage differs from continent to continent, but the trend is the same, with figures varying from 48% to 85%. Urbanization is having a negative impact on society as people gather in cities, resulting in serious traffic congestion if the city does not have public transportation, road networks, and other aspects of a fully developed infrastructure. Another form of impact is air pollution, mostly generated in the transportation sector. In order to reduce such negative impact, it is necessary to plan and implement a variety of measures. One approach is to attract people to live in suburban areas by creating attractive communities. On the other hand, if the trend towards urbanization is inevitable, it is necessary to develop means of transportation that reduce air pollution as much as possible. One approach to reducing air pollution is to convert buses, automobiles, and all other vehicles to 'green mobility', which generates no greenhouse gas emissions. This could involve hydrogen vehicles or pure electric vehicles.

Moving around large cities may still be stressful due to traffic congestion even if 'green mobility' is achieved. Good public transportation networks are based on utilization of railways, trams, buses, and economically shared mobility, with fully developed road networks. If the number of road networks is limited, there could be congestion as trams, buses, and passenger vehicles share the same road. The important thing is to ease traffic congestion and make traffic smoother and less stressful as much as possible.

Easing of traffic requires visualization and control of traffic. Considering that differing local traffic conditions can result in requirements that are unique to the area, the specific approach to visualization and control may differ between countries. Some countries have made large investments in infrastructure and to estimate traffic volume as precisely as possible, while others have not. This difference may originate in the individual countries' concept of how they want to utilize data: controlling traffic based on traffic engineering theory or merely showing the rough traffic congestion. Accordingly, several approaches can be applied, such as detecting traffic volume from roadside units or using in-vehicle GPS data.

There are other elements that affect traffic as well. Pedestrians, motorcycles, and bicycles are major factors that affect traffic. In transit mall areas that do not allow passenger vehicles to enter, but do allow public transportation such as trams, and that have pedestrians walking around the area, the speed of trams and buses differs significantly as the drivers try to avoid hitting pedestrians who are recklessly moving around. On normal roads, motorcycles and bicycles are also sometimes

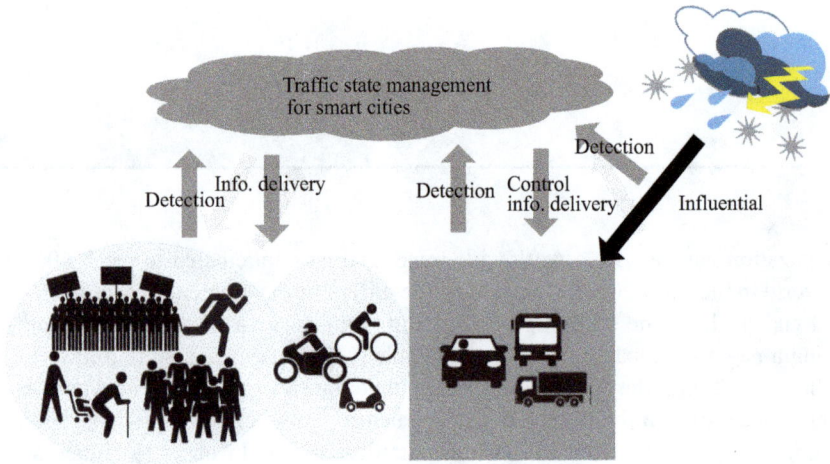

Figure P.1   Holistic approach to detecting traffic states

threats to vehicles, as they may block the vehicle path suddenly, resulting in accidents, or at the very least affect the traffic flow. In addition to these mobile
'players', weather is another factor that affects traffic conditions. Heavy rainfall
may flood roads and prevent vehicle travel.

In order to take a holistic approach to coping with various situations that
require traffic to be monitored and controlled, the various aforementioned factors
that affect traffic must be detected as much as possible. The concept of detection is
shown in Figure P.1. The targets of detection are called 'traffic states' because they
can affect traffic flow in various ways. This book mainly focuses on the detection
side – the detection of traffic states – both from roadside units and on-board units,
in order to examine effective forms of traffic control for the future. There are also
various types of detection unit, such as camera-based devices, millimetre radar,
LiDAR, and GPS-related devices. This book is a collection of information, etc.,
related to the detection of traffic states. Part I explains the perspectives of intelligent infrastructures in three areas: Europe, Japan, and Singapore. Part II demonstrates detection from roadside units utilizing camera-based devices, LiDAR, and
radar technology. Part III addresses detection from on-board units utilizing GPS
and camera-based technology. Part IV mainly explains the detection of vulnerable
road users such as cyclists and crowds of pedestrians. Part V explains the detection
of factors which affect traffic flow. Detection of incidents is one of the two most
significant elements, and detection of rainfall using phased-array weather radar is
the other.

The editor anticipates that readers can utilize the various sensing technologies
to further develop their ideas for future traffic visualization and control.

Part I

Regional activities

Chapter 1

Japan perspective

Koichi Sakai[1]*

1.1 History of intelligent transport system development in Japan

Intelligent transport systems (ITSs) are a new type of transport system that uses state-of-the-art information and communication technologies to integrate people, roads (infrastructure) and vehicles, in order to ensure traffic safety and the smooth flow of traffic and to resolve environmental and other issues. The birth of ITS can be traced back to the 1970s, but it was in the 1990s that the term 'ITS' was defined and came to be used as the common name for this type of system in Japan, the United States and Europe.

In Japan, ITS is expected to make a major contribution to resolving various problems such as traffic accidents and traffic congestion, expanding the markets for automobile and information technology (IT)-related industries and creating new markets. In February 1995, 'Basic Guidelines on the Promotion of an Advanced Information and Telecommunications Society' was accepted by IT Strategic Headquarters headed by the Prime Minister. In August 1995, 'Basic Government Guidelines of Advanced Information and Communications in the Fields of Roads, Traffic and Vehicles' was formulated by the five government ministries and agencies such as National Police Agency, Ministry of International Trade and Industry, Ministry of Transport, Ministry of Posts and Telecommunications and Ministry of Construction. In 1996, 'Comprehensive Plan for ITSs in Japan' was formulated by five government ministries and agencies. This plan is a long-term vision which established the basic approach to target functions, development and deployment of ITS in Japan to be promoted in a systematic and efficient manner, from a user's perspective. This plan established the objectives for research and development efforts by industry, academia and government, organised into 20 ITS user services and 9 development areas (Table 1.1).

In the area of driving assistance, in 1995 the Ministry of Construction conducted automated driving field operational tests and demonstrations on a test course in the city of Tsukuba in Ibaraki Prefecture. The following year, in September 1996, proving tests and demonstrations of automated driving were conducted using the Joshin-etsu Expressway before the expressway was put into service. The system

[1]Institute of Industrial Science, The University of Tokyo, Tokyo, Japan
*Present affiliation: Chiba National Highway Office, Ministry of Land, Infrastructure, Transport and Tourism, Chiba, Japan

Table 1.1   ITS development area and user services in Japan [1]

Development areas	User services
(a) Advances in navigation systems	1. Provision of route guidance traffic information
	2. Provision of destination-related information
(b) Electronic toll collection systems	3. Electronic toll collection
(c) Assistance for safe driving	4. Provision of driving and road condition information
	5. Danger warning
	6. Assistance for driving
	7. Automated highway systems
(d) Optimisation of traffic management	8. Optimisation of traffic flow
	9. Provision of traffic restriction information on incident management
(e) Increasing efficiency in road management	10. Improvement of maintenance operations
	11. Management of special permitted commercial vehicles
	12. Provision of roadway hazard information
(f) Support for public transport	13. Provision of public transport information
	14. Assistance for public transport operations and operations management
(g) Increasing efficiency in commercial vehicles operations	15. Assistance for commercial vehicle operations management
	16. Automated platooning of commercial vehicles
(h) Support for pedestrians	17. Pedestrian route guidance
	18. Vehicle–pedestrian accident avoidance
(i) Support for emergency vehicle operations	19. Automatic emergency notification
	20. Route guidance for emergency vehicles and support for relief activities

was a road-vehicle cooperative system that used magnetic markers and leaky coaxial cable, and automated driving was conducted by platooning 11 passenger vehicles. At that time, this system was referred to as an automated highway system [2].

Subsequently, in September 1996, the Advanced Cruise-Assist Highway System Research Association was founded with automobile manufacturers and electrical appliance manufacturers and the like as members. The focus was changed from automated driving to driving assistance, and the organisation pursued research and development in an effort to develop an advanced cruise-assist highway system based on road-vehicle cooperative systems that use road-vehicle cooperative technologies [2]. This system provides driving assistance by using roadside sensors to detect information that cannot be detected from the vehicle and transmitting the information to the vehicle using vehicle-to-infrastructure (V2I) communication.

Subsequently, a road-vehicle cooperative service using dedicated short-range communication (DSRC) in the 5.8 GHz band was introduced by the Road Bureau of the Ministry of Land, Infrastructure, Transport and Tourism (MLIT) as an electronic toll collection (ETC) 2.0 service. This service was deployed nationwide in 2011, primarily on expressways.

In addition, the development of a driving safety assistance system that will use V2I communication and will be designed primarily for use on ordinary roads is currently underway. This system is being developed as a driving safety support system (DSSS) under the auspices of the National Policy Agency, with development being conducted primarily by the Universal Traffic Management System Society, which has automobile manufacturers and electrical appliance manufacturers and the like as members. Optical beacons and, recently, the 760-MHz band are being used for V2I communication.

The driving safety assistance system using vehicle-to-vehicle (V2V) communication that is discussed in Section 1.5 is being developed primarily by automobile manufacturers as an Advanced Safety Vehicle (ASV) project under the auspices of the Road Transport Bureau of MLIT.

In response to the recent rapid progress in the development of automated driving technologies, since 2014 the IT Strategic Headquarters in the cabinet has been formulating (and updating annually) the Public–Private ITS Initiative and Roadmaps as an integrated comprehensive government policy for promoting measures relating to ITS and automated driving. In addition, the Cross-ministerial Strategic Innovation Promotion Program Innovation of Automated Driving for Universal Services, an automated driving development program of the Japanese government, was initiated by the Cabinet Office to conduct activities that go beyond agency and ministerial boundaries and encompass activities that range from basic research to practical and commercial development.

1.2 Infrastructure sensors and driving assistance using V2I

1.2.1 What is an infrastructure sensor?

In this context, 'infrastructure sensor' can be defined as 'a device that uses infrastructure to detect information and data needed for driving assistance and so on that cannot be detected by the vehicle itself, which are then provided from the infrastructure using V2I communication'. Driving assistance includes driving safety assistance whose aim is to prevent traffic accidents as well as smooth driving assistance whose aim is to prevent traffic congestion. Information obtained from and provided by the infrastructure may include information and data pertaining to signal control, such as when traffic lights are red and when they are green. However, as this information is possessed by the infrastructure itself as opposed to being information detected using the infrastructure, it will be excluded from this definition.

In recent years, the detection technology incorporated in the sensors installed in vehicles has improved, and in many driving environments it is becoming possible to achieve automated driving (including providing driving assistance to drivers) through detection by vehicle sensors alone. However, events that are outside the vehicle sensor's detection range cannot be detected, and it is also difficult for vehicle sensors to detect events ahead that are blocked by physical barriers such as walls and buildings. In such cases, infrastructure sensors that can detect events that are outside the detection range of the vehicle sensor can play an important role.

1.2.2   Events detected by infrastructure sensors

The events that infrastructure sensors are expected to detect can be divided into three general categories:

1. vehicles
2. moving objects that are not vehicles, such as pedestrians and bicycles
3. traffic conditions

Category (1) 'vehicles' can be subdivided into two categories: (a) whether or not a vehicle is present and (b) the status (speed, etc.) of the vehicles that have been detected.

1.2.2.1   Vehicle detection

The detection of vehicles is an issue of particularly crucial importance for assisting driving safety. Most of the various sensors installed in the vehicle itself are for the purposes of detecting other vehicles in the vicinity. In order to avoid collisions with other vehicles, it is necessary

● to detect whether or not there is a vehicle in front when the vehicle in question is moving forward;
● to detect the presence of a vehicle in the other lane when the vehicle in question is changing lanes and
● to detect whether a vehicle is coming from the other direction when the vehicle in question is turning right at an intersection (in a place where vehicles drive on the left side of the street).

In such cases, vehicles at a distance that exceeds the detection range of the vehicle sensor, and vehicles that are in locations outside the detection range of the vehicle sensor because they are blocked by physical objects, cannot be detected by the sensors installed in the vehicle. Using infrastructure sensors to detect vehicles that are outside the vehicle sensor's detection range, and providing that information to the vehicle in question, can help one to ensure driving safety.

In addition, detecting the status of other vehicles in the vicinity such as speed, direction of travel and so on is also an important issue for ensuring driving safety. For example, even if the infrastructure detects the presence of a vehicle travelling in the same direction at a location that is around a curve ahead of vehicle in question and cannot be seen by that vehicle, if that vehicle is travelling at a comparatively high rate of speed (in other words, if traffic is moving), this constitutes normal traffic conditions and will not pose a problem for the vehicle in question, and it will not be particularly important to provide that information to the vehicle in question. However, if the vehicle is travelling at a low rate of speed or is stopped, it may pose a rear-end collision risk for the vehicle in question, and providing the information that has been detected by the infrastructure sensor to the vehicle in question will help one to ensure driving safety.

1.2.2.2 Detection of pedestrians, bicycles and so on

Like the detection of vehicles, the detection of pedestrians, bicycles and so on is an issue of crucial importance for assisting driving safety. The major differences as compared to vehicle detection include the fact that the objects that must be detected are smaller than vehicles and it is difficult to determine whether their direction of movement is one that will pose the risk of collision with the vehicle. Sensors that offer more sophisticated performance will be needed to detect small objects such as pedestrians and bicycles, and so the infrastructure sensors must also be those that offer higher performance. Moreover, pedestrians in particular may not necessarily be moving straight ahead as in the case of vehicles. Therefore, it is necessary to not only detect the presence of pedestrians, bicycles and so on but also to predict their direction of movement. It will be relatively easy to predict the direction of movement if the pedestrians are in a crosswalk or other situation in which their direction of movement is comparatively uniform. However, if there is a pedestrian in an unexpected place, such as jaywalking at a non-intersection location with no traffic signals, it will be relatively difficult to predict the pedestrian's direction of movement, and therefore it will be difficult to determine whether there is a high likelihood of a collision with the vehicle in question.

1.2.2.3 Detection of traffic conditions

Detecting traffic conditions is a special feature of infrastructure sensors. The various sensors installed in vehicles can only detect the presence of a few vehicles in the vicinity of the vehicle in question, as well as their speed and so on. In contrast, detection of traffic conditions by infrastructure sensors constitutes the detection of the general conditions for a certain section of the road ahead of the travelling vehicle (e.g. distinguishing between congestion and heavy traffic) by the infrastructure sensors installed in that location. Vehicle detection as described in Section 1.2.2.1 constitutes only the micro-level perspective: directly detecting 'vehicle presence' and 'vehicle speed' and so on for a single vehicle or a few vehicles. In contrast, detection of traffic conditions constitutes the detection of the average speed and so on for many vehicles and determining the general (average) traffic conditions on the macro-level for the section of road in which the infrastructure sensors are installed, such as distinguishing between congestion and heavy traffic and so on. For example, if the infrastructure sensors detect congestion inside a tunnel (e.g. average speed 20 km/h or less), there may have been a rear-end collision near the tunnel entrance, so providing this information to the vehicle in question will help one to ensure driving safety.

Moreover, detection of traffic conditions is also needed from the standpoint of ensuring the smooth flow of traffic. To take a specific example, if there is heavy traffic ahead in a certain lane on the expressway, even though there is no congestion, something such as a change in incline may result in a decrease in speed on the part of a vehicle ahead that will cause the vehicle behind to brake suddenly, resulting in a deceleration wave or the like that causes traffic congestion. In such situations, smooth traffic flow assistance might be provided, such as recommending to the vehicle in question that it changes to a lane that has less traffic. A particular attribute of infrastructure sensors is this type of traffic flow and density detection for individual lanes, providing a macro-level perspective in which the traffic conditions for many vehicles are detected. This is something that cannot be achieved simply by detecting the status of individual vehicles.

1.2.3   Type of sensors that can be used as infrastructure sensors

1.2.3.1   Vehicle detection

Infrastructure sensors that can be used for vehicle detection include visible light cameras, infrared cameras, laser radar (LIDAR) and traffic counters that detect traffic volume and vehicle speed.

Visible light cameras use images that are the same as the view from the naked eye. Image analysis of these images can be used to distinguish and track vehicles. The main feature of these cameras is their comparatively low cost. However, they tend to be affected by the light environment in the imaging range. For example, if sunlight enters the camera, shadows of structures and the like will be produced in the imaging range, increasing the contrast between the areas in sunlight and shadow and reducing detection performance. Detection performance is also reduced dramatically at night when there are no lights in the area or when visibility is reduced due to fog and so on [3]. In addition, during image analysis, the edges of the vehicle can be distinguished, but, for example if the position at which the camera is mounted is low, multiple vehicles will overlap one another, particularly when there is traffic congestion, with the result that multiple vehicles will be seen erroneously as a single vehicle. In recent years, advances in image analysis technology have helped to resolve these types of issues.

Infrared cameras use infrared rays to make it possible to detect differences in temperature, enabling vehicles to be distinguished and tracked. These cameras are less affected by the light environment than visible light cameras and can provide stable detection performance even in fog or other poor visibility environments [3].

Laser radar (LIDAR) can identify and track vehicles using the principle of determining the distance to an object by measuring the amount of time that a laser beam bouncing off that object takes to be reflected back. The use of a laser beam means that stable detection is possible even at night or under adverse weather conditions.

Traffic counters are mounted on roads by road administrators for the purpose of traffic management. These are infrastructure sensors that detect traffic volume, speed and so on. Formerly traffic counters were loop coil or ultrasonic mechanisms, but in recent years visible light cameras have been used.

Visible light cameras, infrared cameras and laser radar (LIDAR) are capable of identifying each individual vehicle and tracking its movements. They can not only determine the existence of traffic congestion but can also identify whether the end of the congestion is at a particular location. In contrast, traffic counters detect traffic volume and vehicle speed, but as infrastructure sensors they are used to detect the presence or absence of vehicles passing directly beneath that traffic counter.

1.2.3.2   Detection of pedestrians, bicycles and so on

Infrastructure sensors that are used to detect pedestrians, bicycles and so on include laser radar and millimetre wave sensors.

Laser radar is capable of detecting pedestrians, bicycles and other objects that are smaller than a vehicle.

Millimetre wave sensors use millimetre waves with a frequency of 30–300 GHz and a wavelength of 1–10 mm. Millimetre waves follow a very linear path and can be used to detect the distance to the object and its relative velocity simultaneously. In addition, their wavelengths are longer than infrared rays, visible light or radar waves, meaning that stable detection is possible even at night or under adverse weather conditions. They are also capable of detecting pedestrians, bicycles and other objects that are smaller than a vehicle.

1.2.3.3   Detection of traffic conditions

Traffic conditions have been detected using the vehicle identification data (ID), traffic counters and, at the field operational test stage, laser radar.

In the vehicle ID method, vehicle ID collection units are placed at two locations and the difference in the time at which the same vehicle passes by these two units is used to calculate the average speed of travel and determine the congestion status [4]. For example, in tunnels where rear-end collisions occur frequently, the end of the congestion inside the tunnel is constantly changing, and the traffic counter is only able to determine the traffic conditions at the location of that particular traffic counter. The vehicle ID method is effective in cases such as this. In addition, this method is inexpensive, requiring only the placement of vehicle ID collection units. On the other hand, as it measures only the average speed of travel between two locations, it cannot be used to determine detailed traffic conditions.

The traffic counter method detects the traffic volume and vehicle speed in individual lanes and uses this information to determine whether there is traffic congestion at the location of that traffic counter, or whether congestion is likely to occur there [5]. 'Likely to occur' indicates a situation in that the traffic volume is heavy but the speed of travel has not decreased that much. In such situations, if the traffic volume increases even a little bit more, traffic turbulence such as extreme tightening or lengthening of the space between vehicles could cause traffic conditions to change from stable (no congestion) to unstable, ultimately resulting in traffic congestion.

Both the vehicle ID and the traffic counter methods are used to detect traffic congestion. To detect traffic density in each lane, primarily the traffic counter method is used.

1.2.4   *Driving assistance using infrastructure sensors*

1.2.4.1   Driving safety assistance using infrastructure sensors

The following are use cases of the use of infrastructure sensors to provide driving safety assistance.

1. Prevention of rear-end collisions with stopped or slow vehicles ahead on an expressway or at a non-intersection location on an ordinary road, when it is difficult or impossible to see the stopped or slow vehicle from the vehicle in question. Specific examples include preventing a rear-end collision with the end vehicle of traffic congestion around a sharp curve [3]. In such cases, the infrastructure sensor detects a vehicle ahead of the vehicle in question and determines whether it is the end vehicle of traffic congestion. Another example would be in a tunnel or other section where visibility is poor and where rear-end collisions with the end of traffic congestion are likely to occur [4,6]. The infrastructure sensor detects the traffic conditions (vehicle speed) in that section and, if the speed is low, determines the possibility that the end of traffic congestion is present in that section.

2. Prevention of collisions at intersections with a vehicle that is difficult or impossible to see from the vehicle in question. A specific example (in a place where

vehicles drive on the left side of the street) when the vehicle in question is attempting to turn right would be preventing a collision with an approaching vehicle that is attempting to go straight in cases in which there is another approaching vehicle that is also attempting to turn right, and because of that right-turning vehicle the approaching vehicle that is attempting to go straight is difficult or impossible to see from the vehicle in question [7]. In this case, the infrastructure sensor detects the approaching vehicle that is attempting to go straight.

3. Prevention of collisions at intersections with pedestrians or bicycles that are difficult or impossible to see from the vehicle in question. A specific example would be preventing a collision between the vehicle in question, which is attempting to turn right or left at the intersection, and a pedestrian, bicycle and so on in a crosswalk that is difficult or impossible to see from the vehicle in question [7]. In this case, the infrastructure sensor detects the pedestrian or bicycle in the crosswalk at the intersection.

4. Prevention of collisions with a merging vehicle in a merging section, weaving section and so on in an expressway when the merging vehicle is difficult to see from the vehicle in question. A specific example would be a road structure in which the presence of a sound insulation wall or the like makes it difficult to see the traffic conditions in the merging section [8]. In this case, the infrastructure sensor detects the presence of the merging vehicle and provides that information to the vehicle in question to prevent a collision in the merging section.

1.2.4.2   Smooth traffic flow assistance using infrastructure sensors

Infrastructure sensors can help one to ensure the smooth flow of traffic in many ways. For example, at sags, tunnels or other sections where traffic congestion tends to occur, they can detect that the traffic volume is almost at a level at which traffic congestion will occur and then help one to prevent congestion from occurring [5]. In such cases, the infrastructure sensor detects the traffic volume (traffic density) ahead in the lane in which the vehicle in question is travelling and in adjacent lanes.

1.3   Expressway case studies

In Japan, the Road Bureau of MLIT is promoting the development of a 'Smartway' – a next-generation transport system that employs V2I communication technology using 5.8 GHz band DSRC. Field operational tests of driving assistance using infrastructure sensors have been conducted on public roads, primarily on urban expressways.

1.3.1   *Forward obstacle information provision (Sangubashi Curve, Metropolitan Expressway) [3]*

Forward obstacle information provision is a system that uses roadside infra-structure sensors to detect the presence of vehicles that are stopped or travelling at low speed and then uses V2I communication to provide that information to dri-vers. Field operational tests have been conducted since 2005 at the Sangubashi Curve on the inbound lanes of the Metropolitan Expressway Route No. 4 Shinjuku Line.

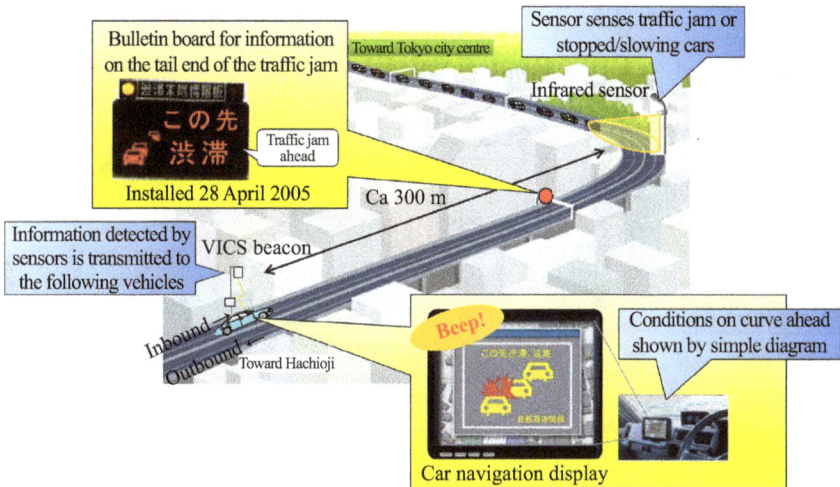

Figure 1.1 Forward obstacle information provision at Sangubashi Curve [9]

The outline is shown in Figure 1.1. The Sangubashi Curve is a sharp curve with a radius of curvature of 88 m. There is a sound insulation wall in this location, and partly for this reason it is not possible to determine traffic conditions around the curve until the vehicle enters the curve. Traffic congestion occurs frequently in the section of road after the curve, and the end of traffic congestion often extends to the area near the curve. There is also a straight section of road approximately 400 m in length immediately prior to the curve, so many vehicles are travelling at high speed when they enter the curve. For these reasons, rear-end collisions frequently occur at the end of traffic congestion in this location.

Four infrared cameras have been installed in the curve section in order to detect the end of traffic congestion. When the infrared cameras detect a vehicle that is stopped or travelling at low speed, this information is provided to the on-board units in vehicles using V2I communication technology from a roadside antenna installed just before the start of the curve. This causes the car navigation system in the vehicle to display a warning to the driver ('Congestion ahead; Drive carefully'), using both images and sounds to urge the driver to exercise caution.

1.3.2 Forward obstacle information provision (Rinkai Fukutoshin Slip Road, Metropolitan Expressway) [6]

At the exit of the Rinkai Fukutoshin (Tokyo Waterfront City) Slip Road on the Metropolitan Expressway Bayshore Route, drivers take an exit slip road from the Metropolitan Expressway main route and then go up a slope, at the top of which is a traffic signal. There is often a line of cars waiting at this traffic signal intersection, and the end of the line of cars at the crest of the slope is difficult to see. Accordingly, there is a potential for rear-end collisions at this location. In 2009, field operational tests of forward obstacle information provision were conducted here.

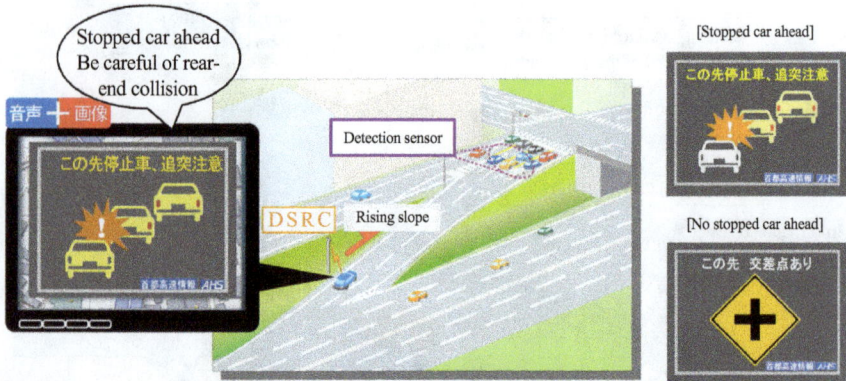

Figure 1.2   Forward obstacle information provision at Rinkai Fukutoshin Slip Road [10]

The outline is shown in Figure 1.2. A visible light camera was installed near the signal intersection to detect the line of vehicles waiting at the traffic signal. When the visible light camera detects a line of vehicles waiting at the signal intersection, this information is provided to the on-board units in vehicles using V2I communication technology from a roadside antenna installed just before the slope. This causes the car navigation system in the vehicle to display a warning to the driver ('Stopped vehicles ahead; Drive carefully'), using both images and sounds to urge the driver to exercise caution. If the visual light camera does not detect a line of vehicles waiting at the signal intersection, the system simply tells the driver that there is an intersection ahead.

1.3.3   Forward obstacle information provision (Akasaka Tunnel, Metropolitan Expressway) [4]

Traffic congestion occurs frequently near the Akasaka Tunnel on the Metropolitan Expressway Route No. 4 Shinjuku Line. For this reason, the end of traffic congestion is frequently present in the tunnel or near the tunnel entrance, and rear-end collisions have occurred frequently at these locations. In 2007, field operational tests of forward obstacle information provision were conducted here.

The outline is shown in Figure 1.3. This location was characterised by the fact that the end of the traffic congestion did not occur in a pinpoint location, such as at the Sangubashi Curve as in Section 1.3.1 or at the signal intersection at the end of the Rinkai Fukutoshin (Tokyo Waterfront City) Slip Road in Section 1.3.2. One method would have been to install a dense array of visible light cameras or other sensors inside the tunnel, but such a system would require a large number of sensors and would be expensive. To provide a simpler, less expensive system to detect the presence of traffic congestion, it was decided to use the method of detecting the average traffic conditions inside the Akasaka Tunnel. Antennas that can use V2I communication technology to acquire ETC IDs were placed near the tunnel entrance and exit, and the duration between the times at which vehicles passed these antennas was used to calculate the speed of travel within the tunnel and determine whether or not there was traffic

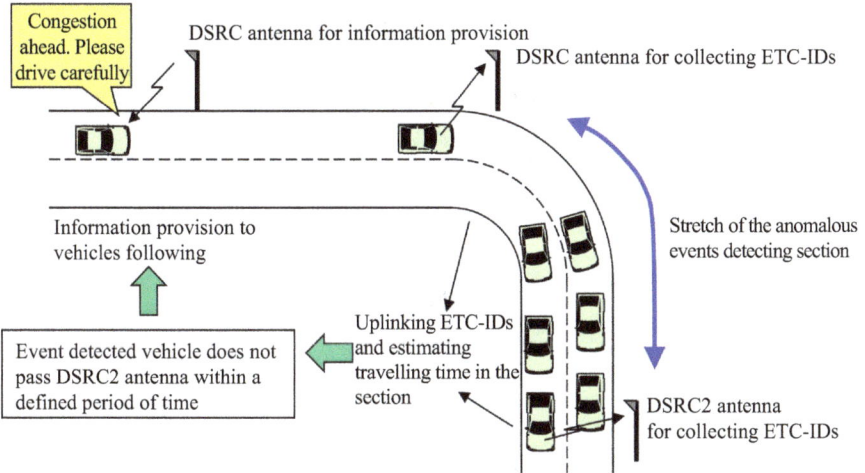

Figure 1.3   Forward obstacle information provision at Akasaka Tunnel [4]

congestion. When the system determines that there is congestion in the tunnel, this information is provided to the on-board units in vehicles using V2I communication technology from a roadside antenna installed at a location before the tunnel entrance. This causes the car navigation system in the vehicle to display a warning to the driver ('Congestion ahead; Drive carefully'), using both images and sounds to urge the driver to exercise caution. In addition to the Akasaka Tunnel, this system has been installed at the Sangubashi Curve and Shinjuku Curve on the Metropolitan Expressway Route No. 4 Shinjuku Line [9].

This system does not use special sensors. It uses the IDs for the ETC system that is widely used throughout Japan and employs simple infrastructure sensors to determine traffic conditions in certain sections of the road. This makes the system comparatively inexpensive, and the roadside antennas that it uses can also be used for other purposes. While the system is not able to determine the exact location of the end of the traffic congestion inside the tunnel, it enhances driving safety by urging drivers to exercise caution before vehicles enter the tunnel.

1.3.4   Merging assistance (Tanimachi Junction, Higashi-Ikebukuro Slip Road and so on, Metropolitan Expressway) [8]

In complicated traffic environments such as merging locations, drivers who are attempting to change lanes need to detect the presence and movements of vehicles both on the road ahead and in the surrounding area. Frequently it is difficult to detect the presence and movements of vehicles in the surrounding area due to heavy traffic or physical objects that obstruct visibility and so on, and as a result, accidents in which one vehicle sideswipes another tend to occur in these locations.

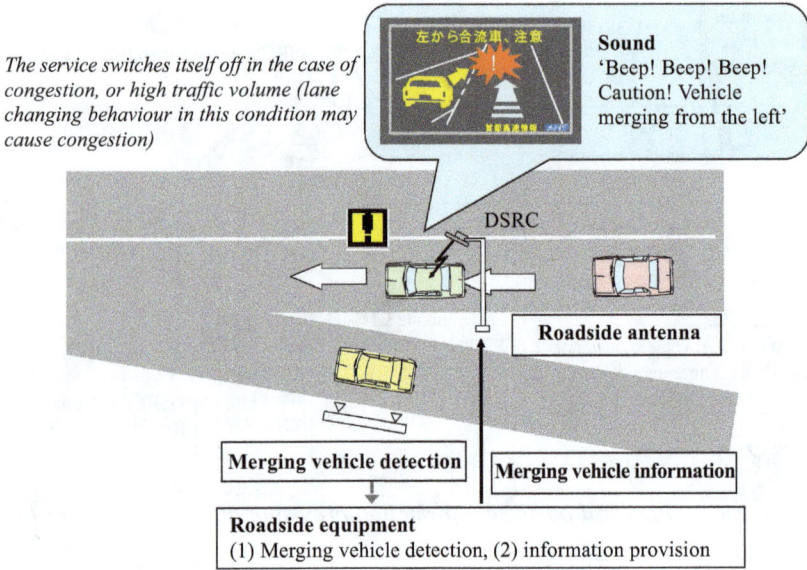

The service switches itself off in the case of congestion, or high traffic volume (lane changing behaviour in this condition may cause congestion)

左から合流車、注意

Sound
'Beep! Beep! Beep! Caution! Vehicle merging from the left'

DSRC

Roadside antenna

Merging vehicle detection

Merging vehicle information

Roadside equipment
(1) Merging vehicle detection, (2) information provision

Figure 1.4   Merging assistance [8]

The outline is shown in Figure 1.4. When physical obstructions impede visibility and it is difficult to see the merging vehicle until just before the merging location, the merging assistance system provides information about the presence of merging vehicles and issues cautionary advisories to vehicles on the main route. Field operational tests have been conducted since 2007 at the Tanimachi Junction (the place where the Inner Circular Route and the Route No. 3 Shibuya Line on the Metropolitan Expressway merge), and at the Higashi-Ikebukuro Entrance Slip Road on the Metropolitan Expressway Route No. 5 Ikebukuro Line.

Merging vehicles are detected by a roadside traffic counter (ultrasonic sensor) placed at the merging lane. This information is provided using V2I communication technology to the vehicle on-board units from a roadside antenna installed on the main route. This causes the car navigation system in the vehicle to display a warning to the driver ('Caution; Merging vehicle from left'), using both images and sounds to urge the driver to exercise caution.

A field operational test was also conducted in 2009 on the Yanagihara Slip Road inbound lanes of the Hanshin Expressway No. 3 Kobe Route [10].

1.3.5   Smooth traffic flow assistance at sags (Yamato Sag, Tomei Expressway) [5]

The widespread use of ETC has dramatically reduced congestion at expressway toll booths. These days, non-intersection road locations – specifically, sags and tunnels – account for most of the traffic congestion that occurs on expressways in Japan. Sags are places where the road incline changes from downhill to uphill. Drivers sometimes do not recognise the change in road incline and so the vehicle speed drops when the vehicle enters the upslope.

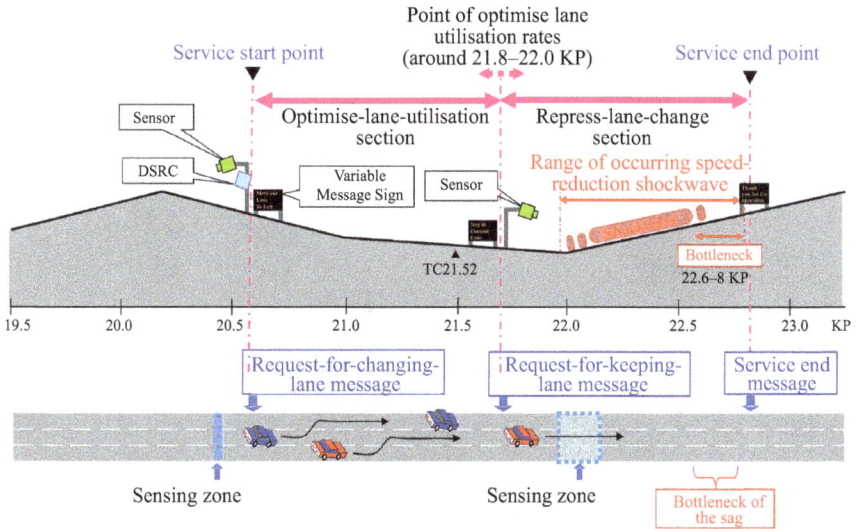

Figure 1.5   Schematic outline of the lane utilisation rate optimisation service at Yamato Sag [5] KP = Kilometer Post

The main cause of congestion at sags is a situation in which there is a high volume of traffic, almost enough to cause traffic congestion with comparatively little inter-vehicular time or headway, and the vehicles ahead slow down due to a change in road incline. This causes the vehicles behind to reduce speed dramatically, producing a deceleration wave. When there is a high volume of traffic, almost enough to cause traffic congestion, in general there is a high volume of traffic in the passing lane (the right-hand lane closest to the median strip). Under these circumstances, if some of the vehicles in the passing lane could be induced to change lanes to the through lane on the left, the average headway between vehicles would become longer, preventing the deceleration wave from occurring.

Based on this concept, a field operational test of a lane usage optimisation service was conducted in 2010 at the Yamato Sag in the inbound lanes of the Tomei Expressway. The outline is shown in Figures 1.5 and 1.6. On the upstream side, information was provided to drivers approximately 1 km before the bottom of the sag (the point at which the road incline changed) urging them to switch lanes from the passing lane to the through lane on the left. On the downstream side, information was provided to drivers approximately 1 km before the bottom of the sag urging them to not change lanes.

Visible light cameras and laser radar were put in place as infrastructure sensors in order to determine the traffic conditions (specifically, the traffic volume and speed) in each lane.

1.4   Case studies on ordinary roads

In Japan, the National Police Agency provides driving safety assistance on ordinary roads. The National Police Agency is working to develop a DSSS to

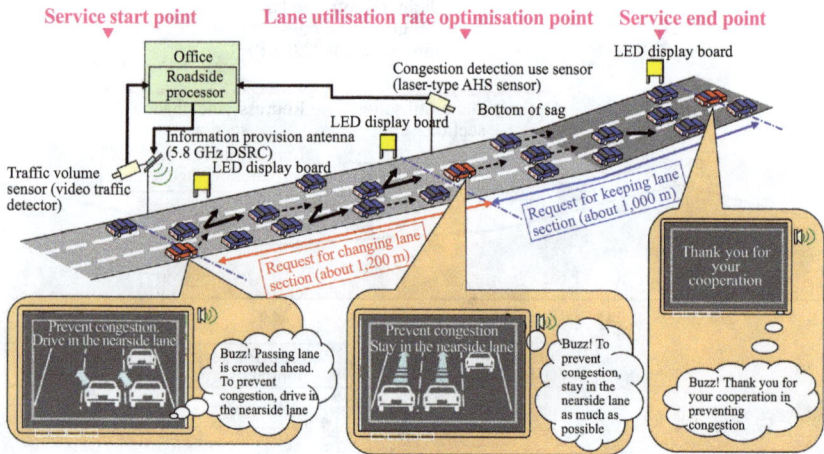

Figure 1.6   Outline of the field operational test at Yamato Sag [5]

reduce traffic accidents. This system will use optical beacons or the 760 MHz band as a V2I communication technology.

1.4.1   Rear-end collision prevention system [7]

This system helps one to prevent rear-end collisions with vehicles stopped or travelling at a low speed due to either traffic congestion or the presence of a stoplight around a curve or up a slope. When the roadside infrastructure sensor detects such vehicles, the information is provided to the vehicle in question via a roadside antenna placed before that section of the road, and the driver is urged to exercise caution.

Field operational tests were conducted, using primarily visible light cameras as infrastructure sensors and using optical beacons for V2I communication.

1.4.2   Crossing collision prevention system [7,11]

This system helps one to prevent collisions at intersections without traffic signals when the intersection is located around a curve or in another location with poor visibility from the main road. The presence of a vehicle on the intersecting side road is detected by a roadside infrastructure sensor placed on the side road and, by means of V2I communication, this information is provided to the vehicle in question from a roadside antenna placed at a location before the intersection on the main road and the driver is urged to exercise caution.

Field operational tests were conducted, using primarily visible light cameras as infrastructure sensors and using optical beacons for V2I communication.

1.4.3   Left-turn collision prevention system [7]

Vehicles drive on the left side of the street in Japan. This system helps one to prevent collisions at intersections when a vehicle is attempting to turn left and a motorcycle or bicycle coming from behind in the left-hand lane is in the driver's blind spot and therefore is difficult for the driver to see. The infrastructure sensor placed at the intersection detects the two-wheeled vehicle, and V2I communication is used to provide this information to the vehicle that is attempting to turn left and the driver is urged to exercise caution.

Field operational tests were conducted, using primarily laser radar and visible light cameras as infrastructure sensors and using optical beacons for V2I communication.

1.4.4   Right-turn collision prevention system [7]

Vehicles drive on the left side of the street in Japan. When the vehicle in question is attempting to turn right, this system helps one to prevent a collision with an approaching vehicle that is attempting to go straight in cases in which there is another approaching vehicle that is also attempting to turn right, and because of that right-turning vehicle the approaching vehicle that is attempting to go straight is difficult or impossible to see from the vehicle in question. An infrastructure sensor at the intersection detects the oncoming vehicle and, by means of V2I communication, this information is provided to the vehicle in question that is attempting to turn right and the driver is urged to exercise caution.

Preparations for putting the system into widespread use are underway, using primarily visible light cameras as infrastructure sensors and using the 760-MHz band for V2I communication.

1.4.5   Crossing pedestrian recognition enhancement system [7]

This system helps one to prevent collisions with pedestrians and bicycles crossing the street at intersections when the vehicle is attempting to turn right, and the pedestrians or bicycles in the crosswalk of the cross street are difficult for the driver to see. An infrastructure sensor placed at the intersection detects the presence of pedestrians or bicycles in the crosswalk and, by means of V2I communication, this information is provided to the vehicle that is attempting to turn right and the driver is urged to exercise caution.

Preparations for putting the system into widespread use are underway, using primarily laser radar as infrastructure sensors and using the 760-MHz band for V2I communication.

1.5   Driving safety assistance using vehicle-to-vehicle (V2V) communication

V2V communication enables vehicles to communicate directly with one another. This enables a vehicle to be alerted to the presence of other vehicles that are difficult for the vehicle to see. Although this cannot be called an infrastructure sensor, it is one method of detecting other vehicles using a means other than the vehicle's own sensors. However, there is an issue in that the vehicle is only able to detect other vehicles that are equipped with V2V communication equipment.

Research and development aimed at finding ways to enhance driving safety using V2V communication is underway by the ASV project being promoted by the Road Transport Bureau of MLIT.

Table 1.2 *Major content of 'ITS-Safety 2010' large-scale proving test conducted in FY 2008 [7]*

		Odaiba, Tokyo	Tochigi pref.	Kanagawa pref.	Aichi pref.	Hiroshima pref.
V2I(DSSS)	Rear-end collision prevention system	○	○	○	○	○
	Crossing collision prevention system	○	○	○	○	○
	Left-turn collision prevention system	○	○			○
	Right-turn collision prevention system	○	○	○	○	○
	Crossing pedestrian recognition enhancement system	○				○
V2V(ASV)	Rear-end collision prevention system	○	○		○	○
	Crossing collision prevention system	○	○			○
	Left-turn collision prevention system	○	○			○
	Right-turn collision prevention system	○	○		○	○

In the 'ITS-Safety 2010' large-scale proving test that was conducted in fiscal year (FY) 2008, many field operational tests were conducted, including tests of both Smartway and DSSS. The details of the content is shown in Table 1.2. Field operational tests of rear-end collision prevention systems, crossing collision prevention systems, left-turn collision prevention systems and right-turn collision prevention systems and so on were conducted in various locations nationwide. The content of each of these systems is generally the same as the DSSS discussed in Section 1.4. In these systems, V2V communication is used to detect vehicles as well as motorcycles and bicycles [7].

References

[1] National Police Agency, Ministry of International Trade and Industry, Ministry of Transport, Ministry of Posts and Telecommunications, and Ministry of Construction (Japan), *Comprehensive Plan for Intellig'nt Transport Systems in Japan*, [online]; 1996, Available from http://www.mlit.go.jp/road/ITS/5Ministries/index.html [Accessed 22 Oct. 2018].

[2] Hosaka, A., Aoki, K., and Tsugawa, S., *Automated Driving – Systems and Technologies*, Tokyo: Morikita Publishing; 2015. pp. 38–9, 129–31 (in Japanese).

[3] Yamada, H., Hirai, S, Makino, H., Yamazaki, T., Mizutani, H, and Oorui, H., 'Development of the Cruise-assist System and Its Effects on the Improvement of Traffic Safety', *Journal of Japan Society of Civil Engineers D*, 2007; Vol. 63 No. 3, 360–78 (in Japanese).

[4] Hatakenaka, H., Sakai, K., Asano, M., *et al.*, 'Proving Tests of the Forward Obstacles Information Provision Service Using the ETC-ID System', presented at ITS World Congress; New York, United States, 2008.

[5] Kanazawa, H., Sakai, K., and Sato, A., 'Proving Test to Develop a Practical Service to Optimize Lane Utilization Rates at Sag Sections of Expressways', presented at ITS World Congress; Busan, Korea, 2010.

[6] Hatakenaka, H., Kanoshima, H., Sakai, K., Asano, M., Yamashita, D., and Mizutani, H., 'Proving Tests in Regional Deployment of Forward Obstacles Information Provision Services and so on', presented at ITS World Congress; Stockholm, Sweden, 2009.

[7] ITS Promotion Committee (Japan), *Press Release: Public demonstration, etc. of a Driving Safety Assistance System Using ITS* [online]; 2008, Available from https://www.kantei.go.jp/jp/singi/it2/others/its_safety2010-syousai.pdf [Accessed 22 Oct. 2018] (in Japanese).

[8] Kanazawa, F., Kanoshima, H., Sakai, K., and Suzuki, K., 'Field Operational Tests of Smartway in Japan', *IATSS Research.* 2010; Vol. 34, pp. 31–4.

[9] Kanazawa, F., Sakai, K., Sawa, J., Ueda, Y., and Morii, N., 'Study of a Method of Quantitatively Evaluating the Effectiveness of Advanced cruise-assist Highway Systems', presented at ITS World Congress; Busan, Korea, 2010.

[10] Fujimoto, A., Kanoshima, H., Sakai, K., and Ogawa, M., 'Nationwide On-Road Trials of Smartway in Japan', presented at ITS World Congress; Stockholm, Sweden, 2009.

[11] Universal Traffic Management System Society of Japan, *Driving Safety Support System (DSSS)* [online]; 2018, Available from http://www.utms.or.jp/japanese/system/dsss.html [Accessed 22 Oct. 2018] (in Japanese).

Chapter 2

European perspective of Cooperative Intelligent Transport Systems

Meng Lu[1], Robbin Blokpoel[1] and Jacint Castells[2]

2.1 Introduction

Intelligent Transport Systems (ITS) for road transport, based on ICT (Information and Communication Technologies), which are rapidly developing around four decades, have the aim to improve traffic safety, traffic efficiency, energy efficiency and user comfort. Core technologies in the ITS domain are sensors, telecommunications, information processing and control engineering. Various technologies can be combined in different ways to create in-vehicle systems ranging from support of singular driving functions to fully automated driving. Systems can be either stand-alone or cooperative, using short-range direct communication for information exchange with other nearby vehicles and the road infrastructure [1].

C-ITS (Cooperative Intelligent Transport Systems), called Connected Vehicles in North America, are a prelude to and pave the way towards road transport automation. Vehicle connectivity and information exchange will be an important asset for future highly automated driving [2]. Infrastructure-based ITS utilise off-board sensors and communication technology. Due to the widespread availability of state-of-the-art communication technology, the deployment of these systems is increasing [3].

This chapter presents C-ITS development and deployment from the European perspective, and is structured as follows: the next section provides a brief overview of the history of C-ITS development and deployment in Europe, through EU (European Union)–funded projects and national initiatives, which are joint research and/or innovation actions of industry partners, authorities and academia, at either a Europe-wide or a national scale. Following sections successively introduce the ITS deployment platform initiated by the European Commission (EC), the C-Roads initiative and main activities of the EU member states, C-ITS architecture, and C-ITS services, use cases and operational guidelines developed in Europe. In the last section, conclusions are drawn.

[1]Dynniq Nederland B.V., Amersfoort, The Netherlands
[2]Applus IDIADA Group, L'Albornar, Santa Oliva, Spain

2.2   C-ITS development and deployment in Europe

C-ITS development in Europe started some 15 years ago in Europe. Under its FP6-IST (The Sixth Framework Programme – Information Society Technologies) funding scheme, the EC launched in 2005 three IPs (Integrated Projects) targeting cooperative systems: SAFESPOT (Co-operative Systems for Road Safety 'Smart Vehicles on Smart Roads', focussing on the in-vehicle side and traffic safety) [4]; CVIS (Cooperative Vehicle Infrastructure Systems, focussing on the infrastructure side and traffic efficiency) [5] and COOPERS (CO-OPerative SystEms for Intelligent Road Safety, focussing on the domain of the road operator) [6]. In 2009, the EU-funded project FREILOT (Urban Freight Energy Efficiency Pilot) [7] was launched, which aimed to develop C-ITS services for freight transport. DRIVE C2X [8], in the period between 2011 and 2014, substantially contributed to the development of V2X (vehicle-to-every-thing) communication technologies. This helped to accelerate cooperative mobility in Europe and enabled a comprehensive Europe-wide assessment of cooperative systems through field operational tests.

The MOBiNET (Europe-Wide Platform for Cooperative Mobility Services) project [9] started in 2012 and aimed to deploy an open platform for offering a solution for a one-stop shop for Europe-wide (roaming and virtual ticketing) mobility services. In 2013–17, the Compass4D (Cooperative Mobility Pilot on Safety and Sustainability Services for Deployment) project piloted three cooperative services (based on IEEE 802.11p), energy-efficient intersections, road hazard warning and red-light violation warning, in seven European cities. The systems are based on a consolidated architecture for optimal interoperability [10]. The German research project CONVERGE (2012–15) created an ITS architecture (called Car2X Systems Network) and mainly focussed on interoperability, economic viability, scalability, decentralisation and security [11]. In the Netherlands, the DITCM program (Dutch ITS Test site for Cooperative Mobility; 2014–15) aimed to accelerate the deployment at large scale of C-ITS and Connected-Automated Driving and developed a reference architecture [12]. Under the umbrella of the Programme of 'Beter Benutten' (Optimising Use), the Netherlands has substantially invested in C-ITS development and deployment. Talking Traffic is a collaboration between the Dutch Ministry of Infrastructure and the Environment, regional and local authorities and national and international companies [13]. This initiative explores new business models and focusses on the following use cases: in-vehicle signage, road hazard warning, priority at traffic lights, traffic lights information, flow optimisation and in-vehicle parking information.

The EU-funded Horizon 2020 project C-MobILE (Accelerating C-ITS Mobility Innovation and depLoyment in Europe) [14], started in June 2017, aims to stimulate large-scale, secure and interoperable C-ITS deployments across Europe and focusses on the deployment of C-ITS services for mobility challenges, including mixed traffic situations in urban areas. The project has defined a set of core C-ITS services needed for Europe, developed related use cases, identified requirements for C-ITS implementation and drafted a procedure for the introduction of C-ITS services in cities. Ex ante cost–benefit analysis (CBA) has been performed. In addition, the project developed a C-ITS architecture, business models for C-ITS and C-ITS implementation guidelines. Within C-MobILE the identified C-ITS services will be fully implemented in nine European cities by 2020, and the impacts will be evaluated.

2.3   European C-ITS platform

The EC Directorate-General Mobility and Transport (DG MOVE) initiated development of a common EU-wide C-ITS platform in 2014 and released a plan for C-ITS deployment in 2016 [15]. The C-ITS deployment platform is intended to play a prominent role in the deployment of automated road transport. After Phase I (2014–16) [15], the resulting shared vision on the interoperable deployment of C-ITS towards cooperative, connected and automated mobility in the EU was further developed in Phase II (2016–17) [16]. The perspective of the C-ITS platform is that ICT infrastructure-based cooperative, connected and automated transport is an option for enhancing traffic safety, traffic efficiency and energy efficiency, and for reducing fuel consumption. The perspective of C-ITS is that ICT infrastructure-based cooperative, connected and automated driving is an option for enhancing traffic safety and traffic efficiency, and for reducing fuel consumption. C-ITS services, determined in Phase I, are presented in Table 2.1. In Phase II, the definition of the services was elaborated in more detail (see Table 2.2).

In 2015, the Cooperative ITS Corridor project started [17], a collaboration between the Netherlands, Germany and Austria, aiming at a joint implementation of C-ITS on motorways in the corridor Rotterdam–Frankfurt–Vienna.

After the launch of C-ITS platform in 2016, a series of C-ITS deployment projects were co-funded by the EC, e.g. CITRUS (C-ITS for Trucks) [18], SolC-ITS (SOLRED C-ITS Monitoring Network) [19], InterCor (Interoperable Corridors) [20], C-The Difference [21] and SCOOP@F [22]. Partners in InterCor are Rijkswaterstaat (Ministry of Infrastructure and Water Management of the Netherlands), the French project SCOOP@F, the UK Department of Transport and the Flanders government in Belgium. In the project, ITS-G5 and/or 3G/4G communication solutions are implemented for operation and evaluation of C-ITS services in the Netherlands, France, the United Kingdom and Belgium. The Cooperative ITS-Corridor project has established a cooperation with the InterCor project. The ongoing NordicWay2 (cooperative, connected and automated transport) project aims to implement C-ITS services in Finland, Sweden, Norway and Denmark using cellular communication (3G and 4G LTE) [23].

Also, in 2016, member states and the EC launched the C-Roads initiative to link C-ITS deployment activities, to jointly develop and share technical specifications, and to verify interoperability through cross-site testing.

2.4   C-Roads initiative

The C-Roads initiative, started in December 2016, with a strong support of the EC, targets interoperability of applications between 16 European member states and takes into account the recommendations of the C-ITS platform. The main focus is on specific testing and deployment activities that support the C-ITS

Table 2.1 C-ITS services in Phase I of the C-ITS platform [15]

Day-1 services	Day-1.5 services
Hazardous location notifications: Slow or stationary vehicle(s) and traffic ahead warning Road works warning Weather conditions Emergency brake light Emergency vehicle approaching Other hazardous notifications *Signage applications:* In-vehicle signage In-vehicle speed limits Signal violation/intersection safety Traffic signal priority request by designated vehicles Green Light Optimal Speed Advisory (GLOSA) Probe vehicle data Shockwave damping	Information on fuelling and charging stations for alternative fuel vehicles Vulnerable road user protection On-street parking management and information Off-street parking information Park and ride information Connected and cooperative navigation into and out of the city (first and last mile, parking, route advice and coordinated traffic lights) Traffic information and smart routing

Table 2.2   Additional C-ITS services in Phase II of the C-ITS platform [16]

Additional C-ITS services	Examples
New additional urban specific services	Access zone mgt. (restricted lanes, zones, tunnels/bridges, mgt. freight loading/unloading areas) V2I Public transport vehicle approaching V2V
Extended functionality of original list of Day-1/1.5 services	Access mgt. of speed – subset: in-vehicle signage V2I On/off-street parking mgt. – subset: on/off-street parking information V2I Temporary traffic light prior. for designated vehicles – subset of traffic light prior. of designated vehicles V2I Collaborative perception of VRUs – subset: VRU protection V2V Collaborative traffic mgt. – subset: connected, cooperative navigation into and out of the city V2I
Additional user groups of existing C-ITS Day-1/1.5 services	GLOSA (Green Light Optimal Speed Advisory) for cyclists V2I

vision relative to traffic safety, traffic efficiency and energy efficiency. Both short-range communication based on IEEE 802.11p (ITS-G5) and mobile cellular communications are leveraged in hybrid communications schemes along around 100,000 km of European roads by C-ITS services. In collaboration with the C2C-CC (Car2Car Communications Consortium), C-Roads has also developed harmonised specifications for four sets of V2I (vehicle-to-infrastructure) applications. These include road works warning, in-vehicle signage, notifications for other types of hazardous locations and GLOSA (Green Light Optimal Speed Advisory). C-ITS has also developed a common European security mechanism for C-ITS [24].

Table 2.3 provides some examples the C-Roads pilot deployment sites throughout the EU member states.

2.5   C-ITS architecture

V2X communication technology is complex. Two main challenges are interoperability and compatibility. It is assumed that both competing access technologies, i.e. IEEE 802.11p WLAN-based and cellular-based communication, will coexist and spread by their own. This also implies that the communication devices have to support both main access technologies, resulting in multi-stack equipment [32].

In Europe, various C-ITS architectures were developed, during the aforementioned projects, e.g. CONVERGE, MOBiNET, Talking Traffic, Compass4D, SCOOP@F and NordicWay. The most recent development within C-Mobile brings architecture one step beyond the current status. Existing architectures are analysed to recognise common concepts and to identify implicit patterns. Based on this, by generalisation of existing C-ITS architectures, a C-ITS reference architecture is created that enables pan-European interoperability of (concrete and implemented) C-ITS architectures. Guided and constrained by this C-ITS reference architecture, a common implementation architecture is defined, specifying components and their interfaces. Figure 2.1 outlines the architecture framework, and Figure 2.2 presents the architecture definition approach for C-ITS [33,34]. The definition approach covers the range from high-level reference architecture via medium-level concrete architecture to low-level implementation architecture. The C-ITS architecture framework outline lists various aspects of the architecture and categorises these into five classes: stakeholders (involved groups), concerns (parameters), viewpoints (perspectives), model types (modelling approaches) and correspondence rules (dependencies of a viewpoint to check completeness and action relationships among the viewpoints to check consistency).

C-MobILE aims to stimulate Europe-wide interoperable and harmonised C-ITS deployment in an open and secure ecosystem, based on different access technologies, the use of which is transparent for service providers, and seamless and continuous for the end users across different transport modes, environments and countries. An open architecture for hybrid C-ITS platform developed is presented in Figure 2.3. Roadside units (RSUs) can be managed with the Cooperative Services Platform (CCSP). The CCSP facilitates the geographical information flow

Table 2.3 Examples the C-Roads pilot deployment sites [25]

Country	C-Roads activities
Germany	Germany is the central partner of the EU-C-ITS Corridor connecting the Netherlands, Germany and Austria [26]. C-ITS Day-1 services are deployed by 17 public–private partners on motorways, and on links to urban areas in the federal states of Lower Saxony and Hessen
Austria	Austria brings in its knowledge gathered in EU-C-ITS Corridor [26] and ECo-AT Living Lab [27]. In 2018, the deployment of the C-ITS infrastructure covering total 300 km of Austrian motorway connecting Vienna and Salzburg was launched. A roll-out is planned in 2019, and additional deployment efforts cover the Brenner corridor and the surroundings of Graz are in preparation
The Netherlands	The Netherlands is one of the drivers of European C-ITS deployment. This is especially visible from the Declaration of Amsterdam [28], signed by European Ministers of Transport under the Dutch Presidency. C-ITS Day-1 services are piloted along the motorway network from Venlo to Rotterdam and the network around Utrecht. An additional focus will be on freight services
Belgium	The deployment efforts focus on the motorway surrounding Antwerp, the motorway connecting Brussels with Aachen (Germany) and Luxembourg, and the motorway towards the Netherlands
France	France has a long C-ITS history. Based on the results of SCOOP@F [29], the pilot deployments are set up across North-East, Centre-East, West, and South-West France, covering the whole functional chain of C-ITS systems and services, aiming to reach a seamless continuity of services at the urban/interurban interface. The deployment includes almost all Day-1 services and part of the Day-1.5 services [27]
Italy	Italian C-ITS deployment focusses on motorways, especially on the Brenner motorway and the motorways close to the city of Venice. The main goal is to implement cooperative systems for the following connected and automated driving applications: truck platooning, passenger cars highway chauffeur, and combined scenarios of trucks and passenger cars. These services will complement the introduction of C-ITS Day-1 services
Spain	Spain focusses on different deployment initiatives. The Spanish Transport Ministry will deploy C-ITS services along the overall road network, covering approximately 12,270 km, using cellular-based communication technology. Additionally more than 380 km will be equipped with C-ITS services based on ITS-G5, covering the metropolitan area of Vigo, the 'Calle 30' in Madrid, and motorways in the north of Spain, and in Catalonia and Andalusia
Portugal	In Portugal the deployment Day-1 and Day-1.5 services covers 460 km of motorway, including cross-border sections in Valença and Caia, and roads giving access to the urban nodes of Lisbon and Porto. The deployment provides services on different kinds of roads (roads in metropolitan areas, interurban roads, streets and highways)
Denmark	The Danish pilot site is part of the NordicWay pilots [30,31] running in the Nordic countries Denmark, Finland, Norway and Sweden. In this pilot, the existing traffic management centre and its backbone systems will provide C-ITS services with a focus on cellular communication technologies

(Continues)

Table 2.3 (Continued)

Country	C-Roads activities
Finland	Finland is part of the NordicWay pilots [30,31] and will operate two major pilot sites. Finnish–Norwegian E8 corridor targets Artic challenge for automated driving in frequent extreme weather conditions (snowy and icy) and low traffic volumes to minimise the safety risks. Scan-Med corridor between Helsinki and Turku, including the urban links, and especially the incident-prone ring roads and arterials in the Helsinki region will deploy C-ITS Day-1 and Day-1.5 services
Norway	Norway is a partner of the NordicWay pilots [30,31]. It provides C-ITS services on the corridor from Tromsø to the Finnish border. The main pilot stretch is from Skibotn to Kilpisjärvi, including road sections on E8. The mapping of the infrastructure for connected and automated driving readiness will start at the E8/finish border and continue south on the E6 to Trondheim and from there to Oslo and onwards to the Swedish border
Sweden	Sweden, as a NordicWay partner [30,31], focusses on cellular communications for C-ITS service provision. Most parts of Sweden will be covered with a focus on the cities of Gothenburg, Stockholm, Södertälje and Uppsala, including their access routes E6, E4, E20, E18 and RV40, which are all part of the Scandinavian–Mediterranean corridor
United Kingdom	The United Kingdom will deploy C-ITS Day-1 services on the A2/M2 Corridor between London and Dover. Both WLAN-based and cellular-based communication technologies will be used along different stretches of the corridor
Slovenia	Slovenia is equipping 100 km of motorway to pilot C-ITS Day-1 services. Its focus of the deployments is the border region with Italy, where severe accidents caused by extreme weather conditions require new services to make Slovenian motorways safer
Hungary	C-ITS deployment started in Hungary in 2015 with the improvement of road safety, especially in work zones, as major aim. A 136-km-long stretch of the M1 motorway between Austria and Budapest was selected for this initial C-ITS services pilot deployment. An extension is undertaken in terms of both geographical coverage and offered services. The focus is on urban deployment, in particular GLOSA and red-light violation. Major parts of the motorway linking Budapest with Slovenia are also in the deployment phase
Czech Republic	In Czech Republic, C-ITS services will be piloted on more than 200 km of motorway connecting Prague with Brno, Hradec Kralove, Pilsen and further towards Germany in the direction of Nuremberg, and towards Austria. Pilot sites in the cities of Pilsen, Brno and Ostrava will verify urban C-ITS use cases with signalised intersections, focussing on level railway crossings

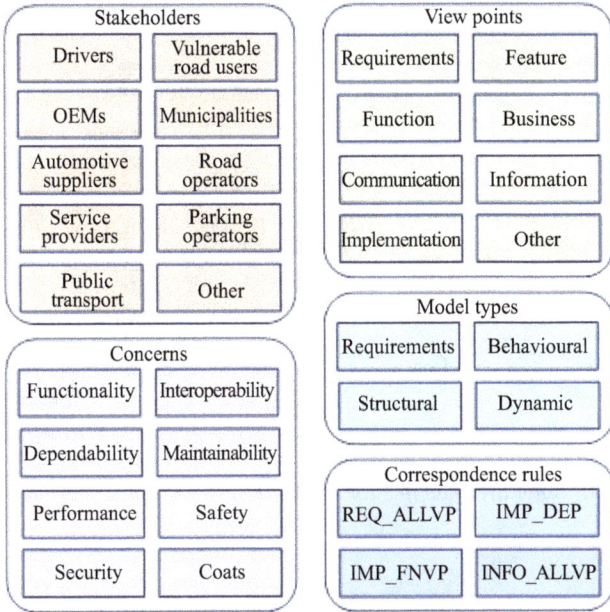

Figure 2.1   Illustration of architecture framework

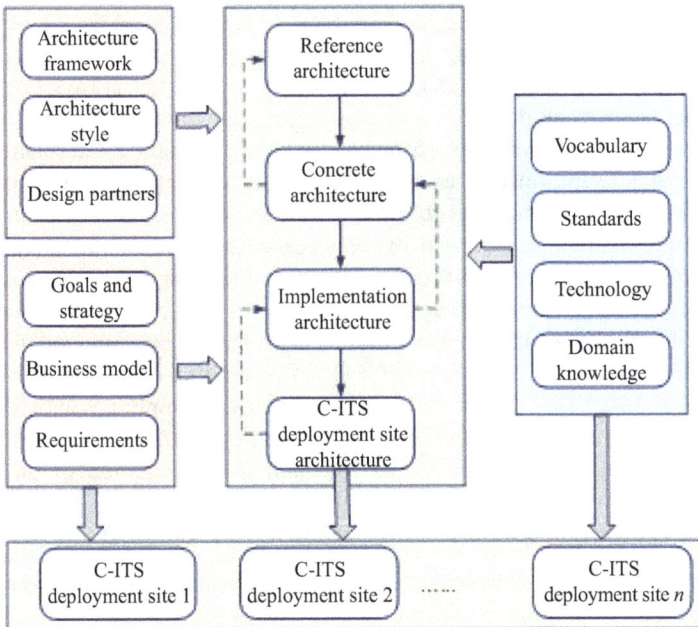

Figure 2.2   Architecture definition approach

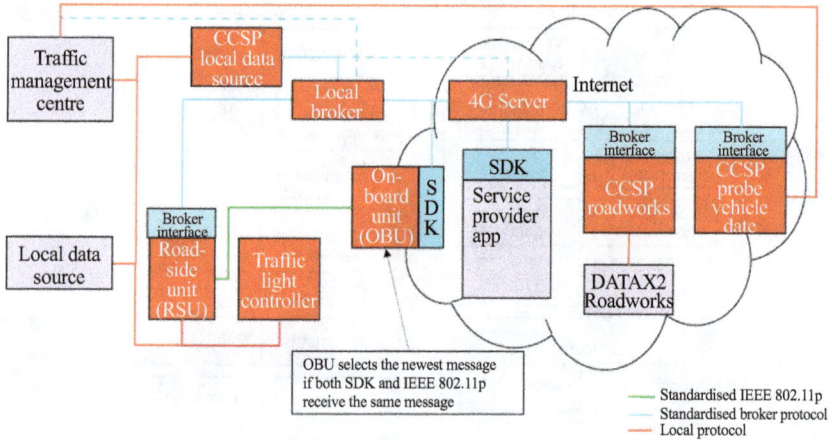

Figure 2.3   An open architecture for hybrid C-ITS platform (adapted after [36])

between RSUs and traffic management centres (TMCs) and enables extension of situational awareness, the cooperative paradigm, beyond just the local dynamic map in a vehicle or an RSU [35].

Offering services via cellular from dedicated C-ITS servers on the Internet starts with data acquisition from local sources. TMCs and local databases are the main candidates for contributing with the necessary data the C-ITS services need, but this could be challenging, as local infrastructure does not always offer interfaces to access relevant data.

Conventional ITS were not so much designed for sharing information with external systems. Information exchange is a core element for providing C-ITS services. Extensive international research initiatives, and efforts of standardisation bodies and legislation, have created, over the past decade, standards and regulations that can today guide and support road operators and local authorities to implement the interfaces that are needed to enable C-ITS.

In addition, the security of such local systems should be evaluated with care, and the large number of data sources for C-ITS, including data from the vehicles on the road, increases the importance of privacy. Complying with the new General Data Privacy Regulation and the public key infrastructure may as well have considerable influence in the design and implementation of the C-ITS architecture.

The proposed European C-ITS architecture has the following merits:

1. It solves the common challenges of secure, private and reliable communication for C-ITS.
2. It provides a standardised mechanism for service delivery of C-ITS applications.
3. It ensures compatibility between existing C-ITS deployments across various cities and regions and serves as baseline for uptake in new locations.

Table 2.4 Main C-ITS services and use cases [37]

No.	Service	Use cases
1	Rest-time management	UC1.1 – Rest-time indication
2	Motorway parking availability	UC2.1 – Information on parking lots location, availability and services via Internet
		UC2.2 – Information on parking lots location, availability and services via I2V
		UC2.3 – Information about a truck parking space released by a user
		UC2.4 – Reservation of a truck parking space released by a user
		UC2.5 – Guide the truck in the port (terminal or truck parking)
3	Urban parking availability	UC3.1 – Information about a truck parking space released by a user
		UC3.2 – Reservation of a truck parking space released by a user
4	Road works warning	UC4.1 – Road works warning for four situations
5	Road hazard warning (incl. jams)	UC5.1 – Hazardous location notification
		UC5.2 – Traffic condition warning, including:
		UC5.3 – Weather condition warning
6	Emergency vehicle warning	UC6.1 – Emergency vehicle warning for three situations
7	Signal violation warning	UC7.1 – Red-light violation warning
8	Warning system for pedestrian	UC8.1 – Safe travelling experience by warning signage
9	Green priority	UC9.1 – Green priority for designated vehicles
10	GLOSA	UC10.1 – Optimised driving experience with GLOSA
11	Cooperative traffic light for pedestrian	UC11.1 – Cooperative traffic light for designated VRUs
		UC11.2 – Cooperative traffic light based on VRU detection
12	Flexible infrastructure (peak-hour lane)	UC12.1 – Flexible infrastructure as in-vehicle signage
13	In-vehicle signage (e.g. dynamic speed limit)	UC13.1 – In-vehicle signage, dynamic traffic signs
		UC13.2 – In-vehicle signage, static traffic signs
14	Mode and trip time advice	UC14.1 – Mode and trip time advice for event visitors
		UC14.2 – Mode and trip time advice for drivers

(Continues)

Table 2.4 (*Continued*)

No.	Service	Use cases
15	Probe Vehicle Data	UC15.1 – Basic probe vehicle data UC15.2 – Extended probe vehicle data
16	Emergency Brake Light	UC16.1 – Emergency electronic brake lights
17	Cooperative (adaptive) cruise control	UC17.1 – CACC passenger vehicles approaching urban environment UC17.2 – CACC passenger vehicles approaching semi-urban environment UC17.3 – Truck platooning UC17.4 – Cooperative adaptive cruise control
18	Slow/stationary vehicle warning	UC18.1 – Slow or stationary vehicle warning
19	Motorcycle approaching indication (incl. other VRUs)	UC19.1 – The approaching two-wheeler warning (V2V) UC19.2 – The approaching two-wheeler warning (V2V and V2I)
20	Blind spot detection/warning (VRUs)	UC20.1 – Digital road safety mirror

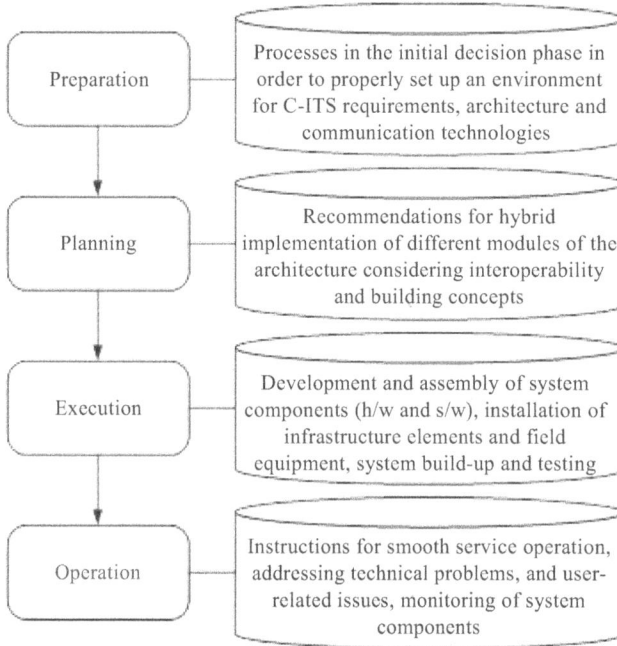

Figure 2.4   Procedure for the introduction of C-ITS services in cities

2.6   C-ITS services and use cases and operational guidelines

Examples of European use cases for the main C-ITS service are summarised in Table 2.4. A survey of the C-ITS services was conducted in 2017, and the results show the importance and potential of the C-ITS services. However, the willingness-to-pay is very low, which underlines the importance of building solid business cases for sustainable large-scale implementation of C-ITS services.

To facilitate C-ITS deployment, operational procedures for introducing C-ITS services in European cities have been defined. Figure 2.4 provides an overview of the approach in a structured procedure of four consecutive phases: preparation, planning, execution and operation. Guidelines for cities for implementing C-ITS services have been developed, especially targeting: road operators, municipalities and technical developers. A detailed description of the operational guidelines for implementation, differentiated for each of the four phases (and as well for some general aspects), is provided in Appendix A [38].

2.7   Conclusions

Interoperable and secure C-ITS applications will make road transport safer, more efficient and more environment-friendly. In the past 15 years, there was substantial activity in Europe to develop C-ITS, and a comprehensive approach was established to investigate possibilities and the best options for C-ITS deployment. The C-Roads initiative brings European authorities and road operators together to harmonise the activities for deployment of C-ITS services across Europe and helps to shape the foundations for Europe-wide seamless deployment of cooperative connected and automated vehicles, and related services.

Guidelines for C-ITS deployment support definition of requirements, identification of deployment locations, implementation of the different components and guidance of end users and management of end-user expectations and issues once the C-ITS services have become operative. Standards, regulations and security are very well considered. Special attention is given to interoperability, and important technical aspects during the implementation phase are explicitly addressed. The guidelines are a useful handbook for cities and technical stakeholders, summarising and applying the current state of C-ITS knowledge, and guiding them in the implementation process. The deployment of C-ITS services will help authorities to facilitate connected, cooperative and automated road transport within a European framework.

The European C-ITS framework has been defined in partnership between major stakeholders and provides a set of key deployment enabling solutions for cities, including related business cases. There are still challenges, related to various aspects: products and services; business development and exploitation; policy, and processes for deploying sustainable C-ITS services, supported by local authorities, and how to ensure interoperability and seamless availability of high-quality services for end users from a perspective of successful business opportunities. Further actions for innovation are (1) assessment, including CBA of the cumulative real-life benefits of clustering C-ITS applications and integrating multiple transport modes in the C-ITS ecosystem; (2) open and secure large-scale C-ITS deployment of new and existing applications that are first demonstrated in complex urban environments and re-interoperable across countries involving large groups of end users; (3) provision of an open platform towards C-ITS sources to support deployment of service concepts for commodity devices and (4) validation of operational procedures for large-scale deployment of sustainable C-ITS services in Europe.

Acknowledgements

The chapter is based on some results of the EU-funded project C-MobILE (Accelerating C-ITS Mobility Innovation and depLoyment in Europe). C-MobILE has received funding from the European Union's Horizon 2020 Research and Innovation Programme under Grant Agreement No 723311. The authors thank the consortium partners, especially Klaas Rozema (Dynniq), Manuel Fünfrocken (University of Applied Sciences in Saarbrücken), Dr. Oktay Türetken (Eindhoven University of Technology) and Dr. Alex Vallejo (IDIADA) for their kind support.

Appendix A   C-ITS implementation guidelines

No.	General issues

G0 A 'city' is defined as all those partners belonging to the decision-making and road management activities for an urban geographical area

G1 A city shall carefully design the activities and timeline involved in the preparation, implementation, deployment and operation and maintenance phases

G2 A city shall identify which technical stakeholders are relevant for the C-ITS deployment and prepare tenders for covering the activities out of its scope

G3 A city shall define its own business model in order to compensate economically, socially, environmentally and/or with safety on roads, the expenses derived from the C-ITS deployment

G4 It is always recommended that experienced stakeholders take specific roles for the reinforcement of availability, integrity, upgrade capacity and quick response

Preparation phase

G5 Cities should define a technical plan for their C-ITS services, which includes identification of achievable technical target, determination of the communication technology, identification of the local C-ITS architecture and identification of the best deployment locations

G6 The requirements elicitation process shall comprise technical and non-technical aspects, in order to cover the entire spectrum of a complex C-ITS deployment

G7 A city shall involve almost all of the partners in the requirements elicitation process to address the different scopes and needs, including end users

G8 There shall be a process of identifying the C-ITS services to be deployed, including the type (V2V, V2I and V2N) which will lead to particular requirements for each service

G9 The requirements shall be revised during the different phases of the deployment in order to properly check if the initial requirements are fulfilled

G10 Cities' main transport objectives are optimal traffic flow, zero road fatalities, emissions reduction, congestion reduction and social inclusiveness, for which the C-ITS deployment shall contribute

G11 Cities should always target a hybrid implementation for their C-ITS services as a future-proof, flexible plan as long as the economic and technical requirements (infrastructure availability, latency requirements) make it feasible and appropriate

G12 The local C-ITS architecture shall be particularly designed taking into account the communication technology used, the C-ITS services to be deployed and the data sources available for them

G13 With an efficient architecture, the ITS-G5 and the cellular-based approaches share the same central system. This provides flexibility and adaptability

G14 A smart dissemination should be offered by the communication provider for enhancing the management of the services and provide hybrid compatibility

G15 For a full integration of the C-ITS services into society, all kinds of users should be targeted, such as pedestrians and vulnerable road users (VRUs), including riders and disabled people

G16 Active detection of VRUs should use the same devices for the notifications to the users when possible (not controlled by the city and very dependent on the scenario the users are located)

G17 Passive detection of VRUs is costly for the city but independent of the communication technology used by the users. It is geographically limited but usually more accurate

G18 Roadside equipment installation shall be carefully chosen in order to avoid malfunctioning, bad services experience and reduced impact of the C-ITS services

G19 The equipment shall be installed where a minimum of information is available, as well as quite close between them in order to properly cover extended areas completely

(Continues)

(Continued)

G20   The antennas should point into the upstream of the vehicles in order to increase the likelihood of a successful communication

G21   Cities and service providers shall be aware that they are receiving 'personal data' from users and they have to comply with GDPR

G22   Cities should explicitly be able to demonstrate users' consent in terms of processing their personal data.

G23   Cities and service providers should take the role of data controllers since they are the entities in charge of the processing and storage of the personal data for the C-ITS services

G24   Cities can allow an external organisation to carry out the processing of the data that they control but do not lose the control of data since they instruct the purpose to that company to process data

G25   Cities need to assign a data protection officer since they process and monitor regular and systemic data from users on a large scale

Planning phase

G26   The role of the service provider is usually being responsible of the provisioning and the software implementation for the different components of the architecture

G27   Every EU deployment city/region shall target pan-European interoperability. The European Commission is making standardisation efforts for the harmonisation of C-ITS deployments

G28   Cities developing C-ITS services shall abide by specified communication standards or definitions for those interoperability interfaces and features

G29   In order to form part of the EC PKI, a RootCA shall be available to provide certificates to the actors of the C-ITS operation. Either the RootCA shall be implemented or used from others' solutions

G30   The TMS are usually connected with the roadside systems and the data providers with 'local' interfaces. These interfaces can remain unchanged for the C-ITS services

G31   A TMS shall either make a decision or adapt and forward the information gathered from roadside systems and data and service providers, depending on the services and the city needs and requirements

G32   The concept of bundling targets a better TMS coordination with some C-ITS services at the same time depending on the current needs, which increases the impact of the services and higher likelihood of success

G33   The TMS shall build a control strategy with the aim of determining which strategy apply in each situation, which are 'inform traffic about the situation', 'enlarge the outflow', 'reduce the inflow' and 'reroute traffic'. Each C-ITS service can be used in some of these strategies

G34   The roadside systems, traffic management system, the data providers and end-user devices are the possible data sources for the C-ITS services. They shall transmit, either directly or indirectly, the information to the service provider

G35   One of the most recommended protocols to transmit information to the service provider is using a REST interface with a PUSH strategy when the information is available in the data sources

G36   The output of the service provider shall be standard C-ITS messages, which are generated based on the information received from the different data sources

G37   The service provider shall generate and encode the C-ITS messages using ASN.1. This means that all entities able to generate and manage C-ITS messages shall implement an ASN.1 encoder/decoder

G38   The service provider shall sign all C-ITS messages for dissemination except for ITS-G5 since the RSU is obliged to sign them too, which would lead to a double security header/signature

(Continues)

(Continued)

G39 The service provider shall either implement or use a dissemination method, which is key for ensure interoperability

G40 The service provider shall be capable of calculating situations/risks for each of the service, in order to properly react to the information received

G41 Road layout information shall be managed by the service provider in order to increase the accuracy and impact of the services

G42 The local service provider shall be the first contact point for the client devices (PID, RSU). It shall implement the interfaces to provide connectivity details based on the location of the devices

G43 The service provider shall implement the capabilities needed for enabling the interoperability, which may require the implementation of standardised interfaces and data formats, depending on the interoperability approach

G44 The communication provider, usually managed by the SP, shall implement a GeoMessaging approach for an efficient hybrid dissemination of the data

G45 Most of the ITS systems (RS) already deployed in a city can be useful for the C-ITS services as well, with little or no adaptation needed

G46 The data acquired from the RS shall be real-time and accurate data for most of the services

G47 The standards implemented by default in ITS-G5 equipment are not enough for covering the different city needs. Some extra implementation is needed mostly in the application layer, for the interaction with the service provider through the communication provider

G48 Knowing the CAN bus architecture of a vehicle and having access to it is not a trivial fact and may cause delays in the implementations and deployments. Ensuring the access to the data beforehand will avoid unexpected issues

G49 Having a GNSS receiver with an integrated inertial measurement unit (IMU) may help to improve the robustness of the information. Moreover, some of the CAN bus data may also be provided, thus avoiding potential CAN access issues

G50 The on-board units (OBU) shall be properly connected with an HMI system to show information to the drivers/end users

G51 Any implementation in the OBUs cannot add high latency to any of the steps for transmission/reception of information

G52 The PIDs shall implement most of the same standards as the RSU/OBUs but taking into account the different communication protocol and the battery consumption

G53 The PIDs shall implement a registration process for getting connectivity details with the proper server, as well as authentication services to use and actions (publish/subscribe) permitted

G54 PIDs shall collect its location information for sending it to the back office in the format required (usually standard CAM messages) or being able to transform it to properly realise the current location

G55 Keep HMI as simple as possible, providing reliable information taking into account traffic conditions and local regulations (e.g. for GLOSA: avoid not applicable/feasible recommendations when traffic jams are present on traffic conditioning vs TTG/TTR prohibition)

G56 When implementing bundling in a city, points of view both end user and operation manager shall be taken into account. These must be integrated into a comprehensive plan that ensures the viability of the implementation

G57 The C-ITS services selection should be based on city/region policy objectives. The most important traffic problems shall be identified and TM strategies shall be created

G58 A city shall identify beforehand the type and number of users of the C-ITS services to be deployed in order to measure the needs of the implementation and ensure a successful operation

(Continues)

(Continued)

G59 In order to ensure the positive impact of the C-ITS services to be deployed, the most critical and relevant parts of the network shall be carefully identified

G60 The involved components shall use a public key infrastructure to be able to trust each other and also other PKI's entities from other regions thanks to a Trust List Manager (TLM) developed by the EC

G61 This PKI infrastructure is intended for all types of devices, including ITS-G5 and cellular-based devices. These devices shall manage special ETSI certificates

G62 CAM and DENM messages are considered personal data and they are sent. The proposed implementation for cellular communication. does not imply sending the location outside the vehicles/VRUs, and the information is always signed so only those authorised entities can receive it

G63 The data controller shall be in charge of ensuring that no personal data is processed and disclosed within its systems

G64 There are five principles for the data processing from the GDPR to follow: purpose limitation, data minimisation, accuracy, storage limitation and integrity and confidentiality

Execution phase

G65 The most recommended height for the RSU installation is 4 m, based on the common installation location of the OBUs in regular cars

G66 The position within the intersection must be as centred as possible in order to provide good coverage for all intersection approaches

G67 The installation must be done by authorised and qualified personnel and must follow the appropriate safety measures

Operation phase

G68 The TMS and the service provider are the main actors involved in the operation of the C-ITS services. This includes the management of the services, actuation plans for dealing with end users and the maintenance of the equipment

G69 During operation, incidences or errors such as broken devices, corrupt data and equipment malfunction may occur. Methods of pre-empting and mitigating incidences or errors must be developed, as well as methods for detecting and correcting them

G70 Each service provider and/or road operator should operate a so-called Pilot Operation and Maintenance Server (POMS) platform, which also helps to monitor the RSUs deployed

G71 It is appropriate to make current and future drivers aware of these technological developments by reaching them through the most appropriate means, be it awareness campaigns or driver training

G72 The traffic management system (TMS) shall be capable of monitoring the entire network constantly and choosing the best C-ITS strategies against the different traffic problems. The service provider shall control the available C-ITS services and manage them following the TMS indications

G73 The development of new systems will be needed to operationally execute the bundling concept. Hence, the existing infrastructure will need to be adapted and new components will have to be integrated

---

References

[1] Lu M. (ed.). *Evaluation of Intelligent Road Transport Systems: Methods and Results*. London: IET (Institution of Engineering and Technology), 2016. DOI: 10.1049/PBTR007E.

[2] Lu M. (ed.). *Cooperative Intelligent Transport Systems: Towards High-Level Automated Driving*. London: IET (Institution of Engineering and Technology), 2019. ISBN: 978-183953-012-8 (Print) / 978-183953-013-5 (eBook).

[3] Zlocki A., Fahrenkrog F., and Lu M. 'History and deployment of ITS technologies'. In: Lu M. (ed.). *Evaluation of Intelligent Transport Systems: Methods and Results*. London: IET (Institution of Engineering and Technology), 2016.

[4] SAFESPOT Consortium. *Technical Annex, SAFESPOT (Co-operative Systems for Road Safety "Smart Vehicles on Smart Roads")*. SAFESPOT Consortium, 2005 (restricted).

[5] CVIS Consortium. *Technical Annex, CVIS (Cooperative Vehicle Infrastructure Systems; Focusing on the Infrastructure Side and Traffic Efficiency)*. CVIS Consortium, 2005 (restricted).

[6] COOPERS Consortium. *Technical Annex, COOPERS (CO-OPerative SystEms for Intelligent Road Safety)*. COOPERS Consortium, 2005 (restricted).

[7] FREILOT Consortium. *Technical Annex, FREILOT (Urban Freight Energy Efficiency Pilot)*. FREILOT Consortium, 2009 (restricted).

[8] DRIVE C2X Consortium. *Technical Annex, DRIVE C2X*. DRIVE C2X Consortium, 2011 (restricted).

[9] Noyer U., Schlaug T., Cercato P., and Mikkelsen L. 'MOBiNET – architecture overview of an innovative platform for European mobility services'. *Proceedings of the World Congress on Intelligent Transport Systems*, Bordeaux, 2015.

[10] Alcaraz G., Tsegay S., Larsson M., *et al. Deliverable D2.2 Overall Reference Architecture (Version 2.0)*. Brussels: Compass4D Consortium, 2015.

[11] CONVERGE Consortium. Deliverable D4.3 Architecture of the Car2X Systems Network. CONVERGE Consortium, 2015.

[12] Van Sambeek M., Turetken O., Ophelders F., *et al. Towards an Architecture for Cooperative ITS Applications in the Netherlands (Version 1.0)*. DITCM (Dutch ITS Test site for Cooperative Mobility), 2015.

[13] Vandenberghe W. *Partnership Talking Traffic*. Dutch Ministry of Infrastructure & Environment, 2017. Available from www.partnershiptalkingtraffic.com [Accessed May 2019].

[14] C-MobILE Consortium. Description of Action, C-MobILE (Accelerating C-ITS Mobility Innovation and depLoyment in Europe). C-MobILE Consortium, 2017 (restricted).

[15] C-ITS Platform. *Platform for the Deployment of Cooperative Intelligent Transport Systems in the EU (E03188) C-ITS Platform Final Report*. Brussels: DG MOVE (Directorate-General for Mobility and Transport), 2016.

[16] C-ITS Platform. *Platform for the Deployment of Cooperative Intelligent Transport Systems in the EU (C-ITS Platform) Phase II Final Report*. Brussels: DG MOVE (Directorate-General for Mobility and Transport), 2017.

[17] Paier A. 'The end-to-end intelligent transport system (ITS) concept in the context of the European cooperative ITS corridor'. *Proceedings of the IEEE MTT-S International Conference on ICMIM*, Heidelberg, 2015.

[18] CITRUS. *2015-BE-TM-0391-S Rhine-Alpine, North Sea Mediterranean Corridor*. CITRUS (C-ITS for Trucks), 2015. Available from www.citrus-project.eu/; https://ec.europa.eu/inea/sites/inea/files/fiche_2015-be-tm-0391-s_final.pdf [Accessed May 2019].

[19] SolC-ITS. *2015-ES-TM-0079-S Mediterranean & Atlantic Corridors*. SolC-ITS (SOLRED C-ITS Monitoring Network), 2015. Available from https://ec.europa.eu/inea/sites/inea/files/fiche_2015-es-tm-0079-s_final.pdf [Accessed May 2019].

[20] InterCor. *2015-EU-TM-0159-S North Sea – Mediterranean Corridor*. InterCor (Interoperable Corridors) Consortium, 2015. Available from intercor-project.eu; ec.europa.eu/inea/sites/inea/files/fiche_2015-eu-tm-0159-s_final.pdf [Accessed May 2019].

[21] C-The Difference. *European Commission N° MOVE/C3/2015-544 Pilot Project: 'Beyond Traffic Jams: Intelligent Integrated Transport Solutions for Road Infrastructure'*. C-The Difference Consortium, 2016. Available from http://c-thedifference.eu/ [Accessed May 2019].

[22] Aniss H. *Overview of an ITS Project: SCOOP@F*, Springer International Publishing, pp. 131–135, 2016.

[23] NordicWay. *2014-EU-TA-0060-S*. NordicWay. 2014. Available from ec.europa.eu/inea/sites/inea/files/fiche_2014-eu-ta-0060-s_final.pdf [Accessed May 2019].

[24] Havinoviski G.N. 'Architecture of Cooperative Intelligent Transport Systems'. In: Lu M. (ed.). *Cooperative Intelligent Transport Systems: Towards High-Level Automated Driving*. London: IET (Institution of Engineering and Technology), 2019.

[25] Böhm M. 'Deployment of C-ITS: a review of global initiatives'. In: Lu M. (ed.). *Cooperative Intelligent Transport Systems: Towards High-Level Automated Driving*. London: IET (Institution of Engineering and Technology), 2019.

[26] BMVI. *Cooperative Traffic Systems – Safe and Intelligent, Introduction to the Corridor: Rotterdam – Frankfurt/M. – Vienna*. Federal Ministry of Transport and Digital Infrastructure (BMVI) Department STB12 Road Transport Telematics, Motorway Service Areas. Cooperative ITS Corridor – Joint Deployment. Available from http://c-its-korridor.de/?menuId=1&sp=en [Accessed May 2019].

[27] ECo-AT. *The Austrian Contribution to the Cooperative ITS Corridor*. ECo-AT Consortium, 2019. Available from www.eco-at.info/home-en.html [Accessed May 2019].

[28] Rijksoverheid. *Declaration of Amsterdam 'Cooperation in the Field of Connected and Automated Driving'*. The Hague: Rijksoverheid, 2016. Available from www.rijksoverheid.nl/documenten/rapporten/2016/04/29/declaration-of-amsterdam-cooperation-in-the-field-of-connected-and-auto-mated-driving [Accessed May 2019].

[29] SCOOP@F. Ministère de l'Environnement, de l'Energie et de la Mer. *Project Scoop – Connected Vehicles and Roads*, 2014. Available from www.scoop.developpement-durable.gouv.fr/en/ [Accessed May 2019].

[30] NordicWay. *NordicWay Evaluation Outcome Report*. NordicWay Consortium, 2017. Available from www.vejdirektoratet.dk/EN/roadsector/

Nordicway/Documents/NordicWay%20Evaluation%20Outcome%20Report %20M_13%20(secured).pdf [Accessed May 2019].

[31]   NordicWay. *NordicWay 2*. NordicWay 2 Consortium, 2019. Available from: www.nordicway.net [Accessed May 2019].

[32]   Jakó Z., Knapp A., Nagy L., and Kovács A. 'Vehicular communication – a technical review'. In: Lu M. (ed.). *Cooperative Intelligent Transport Systems: Towards High-Level Automated Driving*. London: IET (Institution of Engineering and Technology), 2019.

[33]   C-MobILE Consortium. *Deliverable D3.1 High-Level Reference Architecture*. C-MobILE Consortium, 2018 (restricted).

[34]   C-MobILE Consortium. *Deliverable D3.2 Medium-Level Concrete Architecture and Services Definition*. C-MobILE Consortium, 2018 (restricted).

[35]   Dynniq. *V2I (Vehicle-to-Infrastructure) Communication*. Dynniq Nederland B.V., 2015. Available from https://dynniq.com/product-and-services/mobility/cooperative-v2i-services/ [Accessed May 2019].

[36]   Lu M., Blokpoel R., Fünfrocken M., and Castells J. 'Open architecture for internet-based C ITS Services'. *Proceedings of the 21st IEEE International Conference on Intelligent Transportation Systems (ITSC)*, 4–7 November 2018, Hawaii. IEEE Xplore, pp. 7–13.

[37]   C-MobILE Consortium. Deliverable D2.2 Analysis and Determination of Use Cases. C-MobILE Consortium, 2018.

[38]   C-MobILE Consortium. Deliverable D2.4 Operational Procedures Guidelines. C-MobILE Consortium, 2018 (restricted).

Chapter 3

Singapore perspective: smart mobility

Kian-Keong Chin[1]

3.1 Introduction

At the southern tip of the Malay Peninsula and across the Johor Strait lies the island of Singapore. Together with its other smaller islands, Singapore is a city state of only 720 km^2 that housed a resident population of 5.64 million (in 2018). This meant a population density of 7,830 persons/km^2, which is one of the highest ones in the world.

Every weekday, it is estimated that 15.4 million journeys are made in Singapore. During the peak periods of the day, 67% of these journeys (in 2016) are made by public transport. While this is respectable, it does mean that there are still a significant number of journeys made on its road network, which includes 164 km of expressways and 704 km of major roads. These road journeys are contributed by a vehicle population of close to a million (at the end of 1997), of which 547,000 are cars. However, the road network also needs to accommodate other mode of transport of which buses and commercial/goods vehicles are important.

3.2 Challenges and transport strategy

Given the limited land area in Singapore, there is a real constraint on the ability to expand the road network to cope with increasing travel demand contributed by increased human population and economic activities. This means that the transport strategy in Singapore has to favour the use of high-capacity public transport and therefore it is not surprisingly that significant resource and effort are spent in recent years to increase the length of its rail-based mass rapid transit (MRT) or train network. As at the end of 2017, there are 199 km of MRT lines with 119 stations. This is still being expanded and is planned to increase to 360 km by 2030 [1], and by then, it is planned that 75% of daily peak period journeys will be made by public transport. In terms of improving the quality of service, the vision for 2030 is to have 8 in 10 households living within a 10-min walk from an MRT station and for 85% of public transport journeys that are less than 20 km in travelled distance, to be completed within 60 min. However, this increased modal shift to public transport will need concerted effort to achieve, with a transport strategy that is based on a number of push factors even as the improved and yet-affordable public transport system is improved as a pull factor.

[1]Chief Engineer (Road & Traffic), Land Transport Authority, Singapore, Singapore

The push–pull transport strategy, however, has to be formulated in the face of challenges (other than the limited land available for building roads and transport facilities) that Singapore has. While there is still an increasing population (although with controls on immigration, this has moderated in recent times) and hence increasing number of daily trips, the number of elderly has and will continue to increase over the years ahead. This is not expected to reduce the number of daily trips, although it can give rise to changes in travel patterns and travel needs. However, this also means that there will be fewer bus drivers (or bus captains, as they are called locally) and this means that public transport operations must be designed to rely less on drivers. Already, apart from the first two MRT lines, all the others and new ones to be built are or will be driverless. The challenge is also to make buses driverless and this will be re-visited in the later part of this chapter when it touches on autonomous vehicles.

3.3   Demand management – a key element of the transport strategy

Given this difficulty in expanding the road network, it is therefore necessary that demand management be adopted as part of its overall transport strategy. Indeed, congestion management was introduced on Singapore's roads as early as 1975 albeit in the form of a manually operated scheme (and converted to its current electronic or DSRCs (dedicated short-range communications) system in 1998). In 1990, vehicle quota was imposed on the growth of vehicles and while the net growth was 3% at its start, it has been progressively reduced over the years. The latest change in the vehicle quota scheme in Feb 2018 had its net growth slashed to zero (with the exception of commercial vehicle which is maintained at 0.25% annually).

While demand management is to make private transport use less attractive, there is still the need to continue to make travel on the road network manageable and not be stuck with chronic congestion. This is necessary because buses use them, and also freight/logistic vehicles for various commercial activities, and catering for them is necessary, since Singapore is positioned to be a major global business, logistics and financial centre. Singapore has over the years been able to balance this difficult need as evident from global surveys and studies [2,3].

Hence, even with demand management as a key element of the transport strategy, there is a need to introduce traffic schemes to make travel on the roads efficient, safe and sustainable. The use of technology or intelligent transport systems (ITSs) to keep traffic flow respectable is hence pursued with a keen sense of purpose.

3.4   Development of intelligent transport systems in Singapore

The first ITS of many cities would be the traffic signal control system, although it would not have been seen to be so when they were first installed many years ago. The intelligence actually comes when several traffic signals are linked together with a common reference base time so that the operations of these traffic signals are coordinated. This is the same for Singapore when traffic signals in the city centre were linked with telecommunication cables and centrally controlled with a fixed time system in 1981. Different operation plans were used for different time periods of the day and these were pre-programmed based on historical traffic flow patterns. This obviously has the disadvantage of not being able to adapt to changes in traffic flows arising from, say an event or an incident on the roads.

In 1988, this was replaced with an adaptive traffic signal control system. Based on the SCATS system, it was called the GLIDE system locally in Singapore and its settings, and parameters were calibrated to suit Singapore's traffic patterns and operating conditions [4]. The key element of this ITS is the ability to automatically adapt the duration of green times for each approach of junctions and the offsets between sets of traffic signals, based on traffic flow information picked up by loop detectors on the roads. This area-wide adaptive traffic control system allows the following three key functions:

1. dynamically, determine the optimal cycle time for a set of traffic signals and allocate green times to each of the approaches based on competing traffic flows on each of them;
2. determine, at the same time, the offsets between traffic signals to allow green linking so that vehicles have green wave flows on selected routes and
3. since all traffic signals are linked back to the control centre, faults at any of the traffic signals will be picked up almost immediately. This then allows the maintenance crew to get out to the site to restore the traffic signal quickly. With the reduced downtime of the traffic signals, traffic flows much better on the roads.

Following the successful application of the area-wide adaptive traffic control system in the city centre, it was extended subsequently island-wide and today, all 2,450 traffic signals (as at 2017) in Singapore are linked and centrally controlled by the GLIDE system.

Figure 3.1 shows the schematic architecture of the GLIDE traffic signal control system. There are five key elements: the sensors (detector loops), the communications channel, the central computer system or platform, the operating software or applications and the controlled equipment (traffic signals).

In the early 1990s, another ITS to manage the expanding network of expressway was also developed and implemented. This was the Expressway Monitoring and Advisory System (EMAS) which essentially senses traffic conditions and anomalies in the flows, so that the expressway operators can initiate action plans to mitigate and remove the cause(s) of the traffic perturbations. These actions include the activation of emergency vehicles and tow trucks, the dissemination of information to radio stations and social media channels, and to roadside electronic signboards. To ensure that the response time of tow trucks is short enough so that all incidents are attended too, and recovery undertaken to return the traffic condition to normalcy before it gets critical or grid locked, there is a formal contract with a tow truck operator that stipulates response and recovery times for various categories of incidents. Likewise, there is also a formal contract with a number of radio stations to ensure that critical information of incidents are broadcasted to the motoring public in a timely manner.

This system that was initially developed for monitoring and managing traffic flows on the expressways was subsequently extended to cover major arterial roads and major road junctions. With the cameras piping back real-time videos of traffic conditions, the traffic control centre operators would now get an extended view of the situation on the whole road network and allow more effective traffic management.

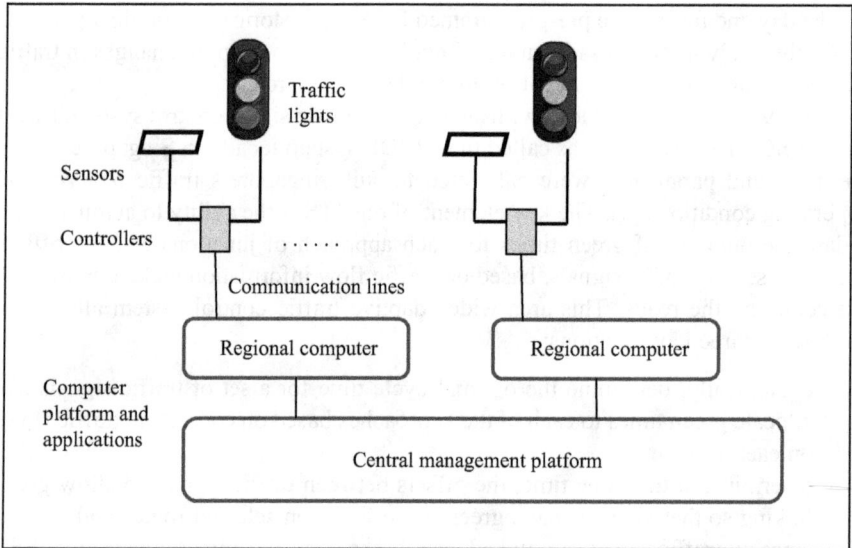

Figure 3.1   Schematic architecture of the GLIDE traffic signal control system

Figure 3.2 shows the schematic architecture of the EMAS. As with the GLIDE system, there are five key elements: the sensors which in this case are mostly video cameras but can include CO_2 detectors and others, for both surveillance and detection, the communication channels which have both optic fibre lines and wireless, the central computer system or platform, the operating software or applications and the controlled equipment (electronic signboards of various types). The computer platform coordinates the operations of the whole system and allows interconnection with other systems, e.g. that of radio stations, social media and emergency response control centres.

The GLIDE system and EMAS make use of fixed sensors installed on or fixed onto the road infrastructure. The development of satellite technology to allow the use of its signals to position itself leads to the development of mobile sensors. These satellite sensors installed onto vehicles are coupled with transmitters to broadcast back to a control centre the whereabouts of those vehicles. This was used initially in the logistics industry and for providing call booking service for taxis. The knowledge of the whereabouts of all unoccupied taxis allows the back-end booking system to assign the taxi closest to the commuter-to-be. This capability was subsequently enhanced with algorithms to compute the travelling speed of the vehicles based on the changing locations of the target vehicle over short duration of time. This was what an ITS system called TrafficScan was set up to do, using location data provided by more than 10,000 taxis on the roads. This large number of taxis provided enough number of data points on major roads to compute traffic condition that can be viewed as a good representation of the actual traffic situation. The value comes from these traffic conditions being transmitted back to control centres for operators to get a sensing of the traffic situation and thereafter, when the same information is broadcasted to the many drivers on the roads. This extra bit of traffic information aids the

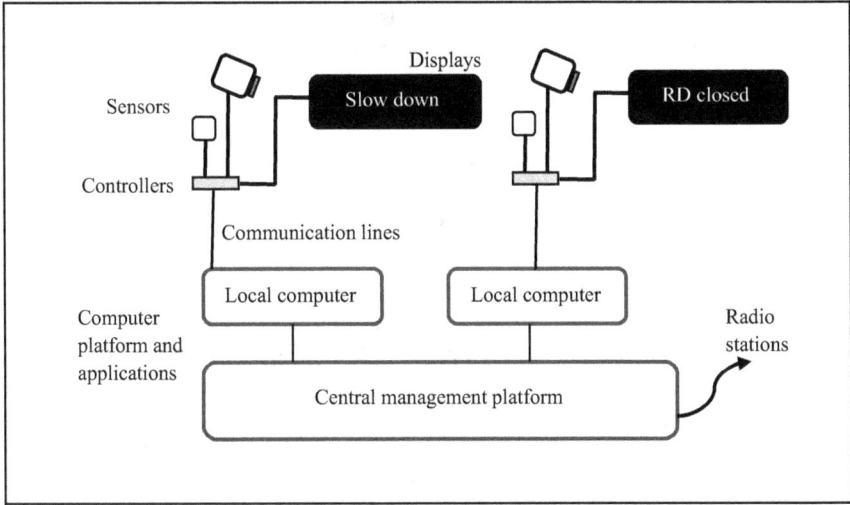

Figure 3.2   Schematic Architecture of the EMAS

drivers in deciding on their choice of routes as the traffic condition on the road changes over time, or in response to an unplanned incident.

With the popularity of smartphones, there are now several applications that harness the location data from these Global Navigation Satellite System (GNSS)-enabled smartphones to compute travel speed conditions on the roads. This method is often referred to as crowdsourcing. In return for providing the location data, the user often gets information on traffic conditions on the roads. It is this ability to get complementary information that has many volunteering to provide their location data. This is necessary as the accuracy of the computed traffic conditions requires a statistically significant and adequate number of data points. Some of these applications have gone another step, in that it can recommend an alternative route that is less congested for the user to reach their destination.

The schematic architecture for using mobile sensors or GNSS-enabled smartphones to crowdsource is similar to that of the GLIDE system and EMAS. It has the same five basic components – the sensors (the GNSS-enabled devices), the communication channels (cellular wireless network), the back-end computer system or platform, the applications that do the computation of average travel speeds and traffic conditions, and the devices on which the useful traffic information is conveyed to (which are usually the same smartphones that provided location and related data).

Several smaller ITSs were also introduced over the years that do the job of monitoring traffic devices, particularly for fault reporting so that rectification works can be initiated quickly. These include variable traffic signs, whose display can be changed or removed based on time of day or on occurrence of events. These displays can show that restriction of travel speeds is in force at school zones at the start and end of school hours, the ban on specified turning movements at junctions during peak traffic periods, and as alerts when vehicles are detected travelling above the stipulated speed limit at tight bends along roads. These ITSs were wired back to the traffic control centre to allow the monitoring of the many traffic devices in use.

3.5   Integrating ITS on a common platform

With the development and implementation of the several ITSs over many years, each being independent and discrete systems, it became obvious that some form of integration was necessary. Given their being separate systems, each operator would have several monitoring displays and keyboards on their workstations – one set for each ITS. It was troublesome for the operators to manage an incident that required manipulation of controls of a few systems at the same time. For example, the operators may need to manage the GLIDE system as well as the EMAS if an incident occurs on the exit ramp of an expressway where there is a set of traffic signals where it joins the local road network.

Hence, in the mid-2000s, several operating ITSs were brought together under a common platform which we call the i-transport platform. One of the initial activities for this exercise was to standardise data exchange formats so that when these ITSs were on the common platform, commands executed by a single keyboard could be understood by all the systems, and also to facilitate the automatic transfer of data from one system to another. This decluttered the operators' workstation significantly and also allows information of operating status of several ITSs to be displayed on the same computer display terminal, although it might still be on individual windows. The resulting system (i-transport) is actually an Internet-of-Things (IoT) setup, and its schematic architecture can be represented in Figure 3.3.

In this schematic architecture, there are again the five basic components – the sensors, which now come from all the constituent systems, the various communication channels, the applications that drive all the functions, the common platform that integrates the flow of data and controls and the various controlled devices or displays of output from the various systems. The applications can reside in their individual ITS or it can be on the common platform and this is more likely the case if the application uses data from more than one ITS or controls more than one ITS.

The i-transport project was, however, more than just an integration exercise to bring several ITSs under a common platform. It was also to bring improved operating processes and capabilities to the operations of traffic management. One of these improvements was the setting up of a local library of display messages that can be displayed on electronic roadside signs. This library was initially populated with often-use messages but has the capability of new messages to be developed and added over time. Another useful feature was the introduction of intelligence with an application into this library. This application would recommend the most appropriate messages based on the scenario sensed by the inputs from the many sensors in use. This reduced the effort of the operators who would no longer need to figure out what the situation was based on his manual interpretation of the several pieces of data received, and then creating the most appropriate message. Instead, the operator only needs to confirm and accept the recommended message from the library. The operator would still have to option to reject the recommended message, following which the application would recommend another, or the operator can create an entirely new one.

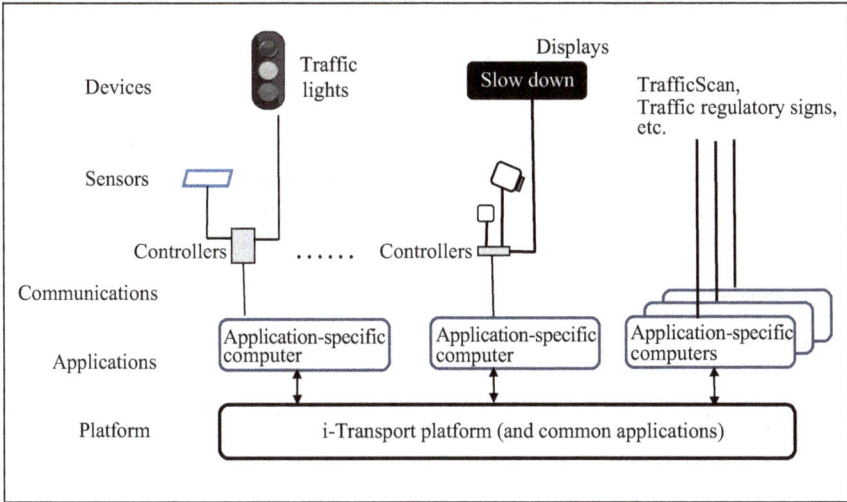

Figure 3.3   Schematic architecture of i-transport as an IoT setup

Another was the development of a common data hub where relevant data from several sources or ITSs are stored. The types and format of the data stored were worked out earlier with various users, although there is an option to include more data types and format when new sensors or new needs arise from various users. The data can be retrieved by the users based on templates set up earlier and can also be exported into common data analysis applications for further analysis by various users.

The ITSs described so far are to optimise available road capacity in the road network. For example, GLIDE optimises traffic signal timings and linking along selected routes to maximise flows, while EMAS and TrafficScan allow any incident to be detected and removed quickly so that flows return to its optimised state as soon as possible. EMAS and TrafficScan also provide information to road users so that they may decide on the best route for them to reach their destination quickly. The information can be used before the start of the trip as in route planning, or en-route as in diverting to an alternative should the original route becomes more congested than anticipated.

However, another aspect of managing traffic flows and making it optimised lies in the application of demand management. Demand management essentially reduces the demand for travel on the road network by, for example diverting some of these trips to public transport (which is more efficient in that it carries more people for the same road space used) or by diverting them to less congested roads or less congested time periods. Road pricing is one of the tools that have been used in Singapore and ITS can play a role in this demand management strategy.

3.6   Road pricing in Singapore

3.6.1   *The manually operated Area Licensing Scheme*

Singapore has a long history in road pricing and has the first practical application way back in 1975. It was essentially a manually operated scheme given the state of the technology at that time, and its key purpose was to manage congestion on roads leading into the central business district (CBD). Called the Area Licensing Scheme (ALS) [5], it initially required all cars to have valid paper permits displayed on its windscreen to enter the cordon that surrounded the CBD. Subsequently, this scheme was extended to other types of vehicles, i.e. light goods vehicles and motorcycles and also expanded to include congested stretches of the expressway network outside the CBD. These daily permits were sold at kiosks located outside the CBD, post offices and convenience stores. For regular users, monthly permits were available for them at post offices, with a discount given so that it's cost was equivalent to 20 days of usage for that month. These permits were valid throughout that day or month, that is the vehicle would be able to enter the CBD for an unlimited number of times for that day or month.

To ensure effectiveness and compliance, enforcement officers were stationed at all pricing points on the cordon or expressways. They were to check that all applicable vehicles have valid permits displayed, and those without valid permits displayed had their vehicle's registration licence number noted down together with other key identifiers, e.g. colour of the vehicle. The owner of that offending vehicle would subsequently be issued with a summons.

As would be obvious with this manual process, there were challenges. For one, it was unlikely that the enforcement officers stationed at the pricing points would be able to spot all offending vehicles, particularly when there was heavy rain. However, it did not matter if enforcement was not 100% as the key outcome was to have an effective scheme. Indeed, it had been effective as a study done by the World Bank [6] had shown that the number of cars entering the CBD fell by as much as 73%.

Another challenge was that while the permits were non-transferable, it would be difficult to spot this based on the vehicle's registration licence number written on the permits. Another criticism of this manually operated scheme was that the occasional motorists entering the CBD would have to detour to a sales outlet to buy the permits before entering the CBD.

A further challenge was that, given its labour-intensive operations, there was difficulty in recruiting the large number of personnel to operate the scheme, especially as the number of pricing points increased with more traffic and congestion on the road network.

3.6.2   *Road pricing adopts intelligent technologies*

Given the challenges with the manually operated ALS, it was inevitable that it migrated to an electronic system, using then state-of-the-art ITS. This was a multi-lane free-flow scheme that used DSRC technologies to communicate with the vehicles and undertook various functions, including that of deducting road pricing charges for the use of the road. At all pricing points, there were custom-built charging gantries with antennae and enforcement cameras to sense and communicate with electronic in-vehicle units (IUs) installed inside vehicles. Called the electronic road pricing (ERP) system [5,7], this overcame some of the challenges that were experienced with the manually operated ALS.

The technologies adopted for the ERP system can be classified into several components, as shown in Figure 3.4. These are the vehicle identifiers in the form of an electronic tag called the IU, the stored-value smart card or credit card, the DSRC communication system, the on-site equipment to detect and deduct the appropriate

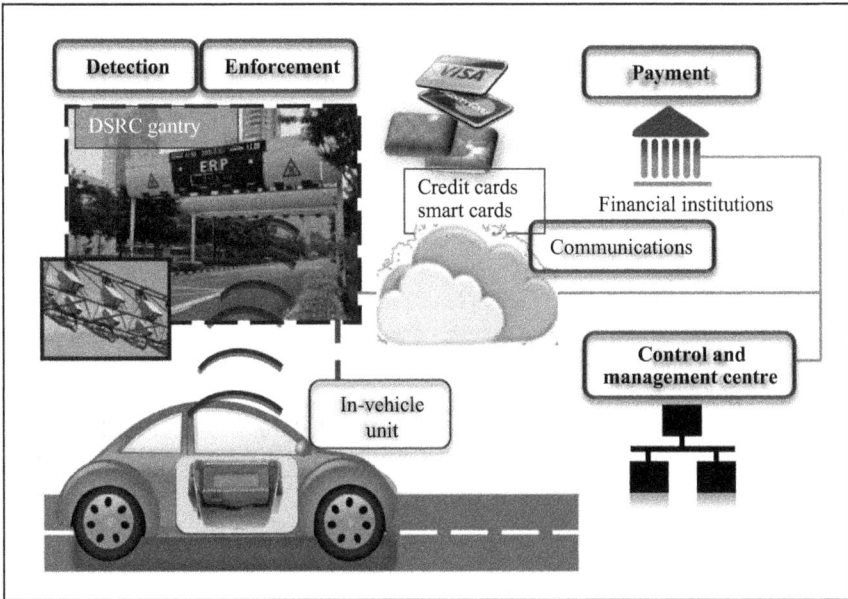

Figure 3.4   Schematic of the various components in the ERP system (Source: Land Transport Authority, Singapore (2017))

road pricing charge, the camera-based enforcement system and the back-end Control and Management Centre.

The ERP system did away with the single-fee unlimited entries into the CBD for the whole day. Instead, it became a pay-per-pass system which was more equitable, since the heavier users of high-demand road space were now paying more road pricing charges. This is especially true since the level of pricing charge varies based on the level of congestion by time and by location. Different pricing levels for different vehicle types are also used, on the basis that larger vehicles occupy more road space and were charged more for each passing. For example, large heavy vehicles were charged at twice the rate of that payable by cars and cars, in turn, pay twice that of motorcycles.

The road pricing charges are based on an electronic rates table stored in the back-end computers and uploaded to the pricing gantries. This means that the road pricing charges can be fairly easy to amend, and this was the case as traffic conditions are monitored at all pricing points so that the road pricing charges can be amended based on observed traffic conditions. Traffic conditions on the roads are measured using average travel speeds as proxies, and this was done every 3 months. On expressways, the acceptable speed range is 45–65 km/h, while for other roads, this is 20–30 km/h. Average speeds below these ranges meant that traffic conditions have worsened enough to warrant increases in road pricing charges. Conversely, if average speeds go above the ranges, road pricing charges are adjusted downwards. The acceptable speed ranges were derived using empirical studies on speed-flow relationships and a wide enough range was selected to optimise flows without constant oscillation in pricing levels. During the long school holidays in June and at the end of the year, some of the pricing points also have their pricing levels lowered (or removed) on the basis of expected lower traffic flows.

Enforcement was almost foolproof with the enforcement cameras at all pricing points, although there is still an appeal process for motorists issued with summonses. In addition, the penalty for non-payment is not viewed as an offence and the follow-up action is by way of paying the applicable road pricing charge with an administrative fee (for the manual effort in processing the road pricing charge payment). To drive home this point, payment via on-line channels carries a lower administrative fee, since these require lower manual effort.

Payment of road pricing charges is through the deduction of stored cash value in CEPAS[*] smart cards. These are readily available and can be topped up easily, up to a value of S$500.[†] Hence, this does away with the need for motorists to detour from their planned route to buy paper permits under the ALS. Subsequently, an added payment convenience was to have these made through credit card payments at the end of each month, thereby doing away with the need to have a stored value CEPAS smart card inserted into the IUs.

3.6.3   Challenges with the ERP system

The DSRC-based road pricing system with its pricing gantries is still a point-based charging scheme. This can give rise to the situation where motorists who travelled on the highly trafficked stretch of roads upstream of the pricing gantry are not charged if their destination is just upstream of that pricing gantry. Similarly, motorists whose destinations are just downstream of the pricing gantry are charged the full pricing charge as those who travel substantially beyond that pricing gantry. Hence, this can be viewed as not being equitable. Instead, the pricing charge paid by each individual motorist should correspond closer to the actual distance travelled on priced roads.

The pricing gantries are heavy infrastructure since these are installed with many electronic devices and components. Each has a clear headroom of at least 5.6 m to avoid any vehicles hitting the many electronic components installed on them. This means that the cost of each pricing gantry is substantial and any relocation due to development works or due to changing travel patterns on the road network would be expensive. Moreover, given that most of Singapore's utility cables and pipes are laid under the roads, there may be difficulty in placing the pricing gantries at their most optimal positions in all instances. This resulted in some of the pricing points being second best options and therefore less effective as a traffic management tool.

3.6.4   The next-generation road pricing system

The realisation of these challenges spurred the development of a new system that does not need such heavy infrastructure. This new system will use GNSS technology to provide position of the vehicles, instead of relying on physical pricing gantries. This is the key aspect of the on-going contract to upgrade the ERP system.

The determination of the exact position of vehicles will be performed by smart applications in electronic on-board units (OBUs) using satellite signals received

---

[*]CEPAS (Contactless e-Purse Application Specification) is a Singapore standard for stored-value smart cards.
[†]US$1.00 = S$1.38 (as at Oct 2018).

from three satellite constellations. There are integrated chipsets available in the market to handle satellite signals from GPS, GLONASS and BeiDou. Where satellite signals may be weak as in areas with 'urban canyons' in the densely built-up city centre of Singapore, the use of augmentation beacons will be introduced. These beacons would use DSRC at 5.9 GHz to allow the OBUs to determine its locations when the satellite signals are unable to. There will be safeguards built into the system to make the OBUs resistant to jamming and spoofing. In addition, all communications are designed to be secured using appropriate encryption and verification techniques, and the OBUs will have anti-tampering features built-in as well.

The road pricing application in the OBU will monitor the position of the vehicle, and when it is sensed to be within a defined zone that needed the payment of a road pricing charge, instructions will be issued to either deduct it from a stored-value smart card or have the deduction made from a designated bank or credit card account as back-end transactions. Using satellite signals to determine the vehicle's location makes possible the use of distance-based congestion charging on the road network, although point-based pricing, either as part of a cordon or individually, will continue to be a feature for effective traffic management on Singapore's urban road network. However, when this next-generation road pricing system is introduced sometime after 2020, it is expected to retain the same pricing points and pricing structure as the current ERP system (Figure 3.5).

While payment of the road pricing charges can be from stored-value smart cards or from a back-end account opened by the vehicle owner or user, the latter is preferred. This is because smart-card-based transactions tend to result in 'non-compliance' when users forget to reinsert the stored-value smart card after taking them out for a number of reasons such as for topping up its balance. Their back-end account will be automatically topped up with a fixed amount from their designated bank account each time the balance drops below a pre-determined threshold. Alternatively, the back-end account can be linked to a credit card, and road pricing expenditure is paid together with other credit card expenditures at the end of each billing cycle (usually a month).

Ensuring that motorists comply and pay with this GNSS-based system is still necessary, and this can be achieved with the continued use of enforcement cameras. These cameras will have complementary beacons to interrogate vehicles to ensure that there are working OBUs on board them. In the event that there are non-compliance issues, the cameras will be activated to capture the vehicle's licence plate and enforcement action can follow thereafter. These cameras can be located at pricing points or along pricing routes (for distance-based pricing), but the roadside infrastructure associated with these cameras is relatively less heavy. Moreover, enforcement can also be made with portable cameras that can be moved around the road network, and with mobile cameras installed on moving vehicles (Figure 3.5).

As with the TrafficScan system that relies on the GNSS-equipped devices on taxis that serve as vehicle probes for determining travel speeds on the roads, the GNSS-equipped OBUs on vehicles will also be capable of providing this same traffic information. With a significantly larger pool of vehicles to rely on, the real-time information on the congestion levels on the stretch of roads that they are on will likely to be more comprehensive and reliable. Data from only a sample of vehicles is needed and these can be analysed, in an aggregated and anonymised manner, to provide traffic conditions on the road network. The information on traffic conditions for roads in the network are then disseminated in real time to

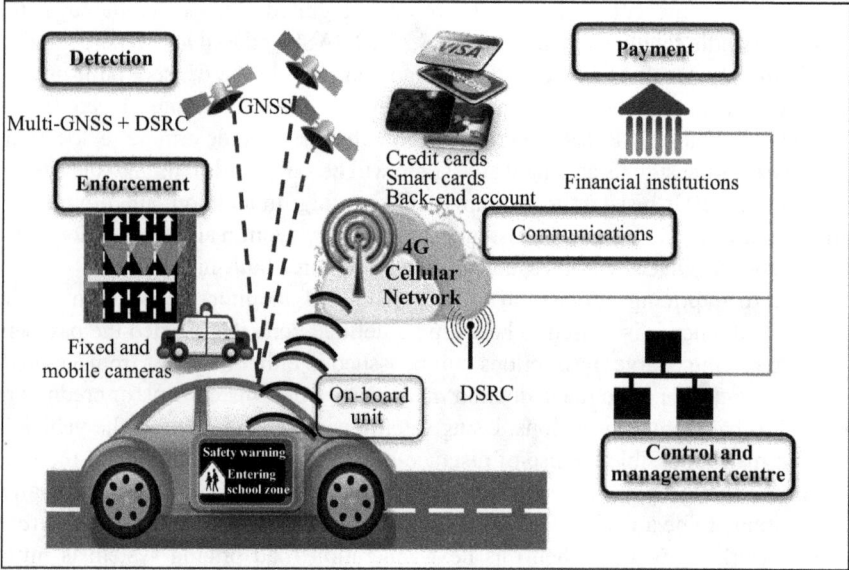

*Figure 3.5   Schematic of the various components in the next-generation road
pricing system (Source: Land Transport Authority, Singapore (2017))*

motorists, either via the installed OBUs or via other electronic platforms, e.g.
smartphones or roadside electronic display panels. Motorists will hence be able to
make pre-trip planning on their travel routes as well as make adjustments to the
travel routes when on the roads. This will especially be useful when there are ad-
hoc incidents that caused unanticipated congestion on the roads.

Other types of useful information can also be disseminated via the OBUs. For
example, information on events that can potentially cause traffic congestion on that
day (or any other days) can be displayed on the OBUs to motorists when they start
their journey (so that they can make adjustments to their travel plans). Depending
on arrangements made with parking garages, the availability of real-time parking
spaces can also be displayed when the vehicles are in their vicinity. In this manner,
motorists can be diverted easily to alternative parking spaces when their earlier
preferred one is shown as being fully occupied.

Other applications with the OBUs include the convenience of serving as an
access card for secured entry into specified parking garages and also allow the
payments of parking charges conveniently through the OBUs. Roads or areas off
limit to special vehicles, for example those carrying hazardous materials, can be
identified in the OBUs and drivers of these alerted when they come within their
proximity. These alerts can also advise them of alternative routes to take. It is also
envisaged that should the driver not heed the advice (or instruction) to use the
alternative route, the traffic operation control centre can be alerted as well and the
vehicle intercepted before it reaches the off-limit roads or areas.

3.7 Big data and analytics for traffic management and travellers

ITSs have been relied on in the past many years to improve efficiency and safety in our transport systems as it became evident that urban cities have limitations to build up more capacity in their infrastructure. Primarily, the limitation for Singapore has been in the availability of land which meant that new physical transport infrastructure has to be built predominately underground. This includes major roads and also the rail-based MRT system.

Traffic lights have been integrated and operated intelligently with an area-wide coordinated system. Road operations have also become more efficient with more sensors and cameras to give operators an integrated overview of the traffic chokepoints and therefore the ability to shorten detection and recovery processes of incidents. Demand management has also become more targeted and effective with the ERP system. However, all these ITSs were to make the use of infrastructure more efficient, as well as allowing the operators to become more effective in their traffic management role.

While the ERP has also influenced the behaviour of motorists to the extent of making them change their travel habits to further optimise the use of the road network, there is the realisation that more can be achieved with ITS to more effectively influence motorists' behaviour. This can come from more intelligent analysis of traffic data, and the dissemination of the analysed information in a manner that can improve the influence on motorists.

Indeed, in the 2014 ITS Strategic Plan for Singapore [8], which is aptly titled 'Smart Mobility 2030', the focus was on the role of traffic data and information for a more connected and interactive land transport community. The connectedness is not just among road users but also with infrastructure and vehicles, including connecting vehicles among themselves.

3.7.1   Quality of data and information

The ability to make intelligent use of data must be on the premise that there is good-quality data. This, in turn, means that there must be dependable and comprehensive network of sensors on the road network and this is becoming mainstream as cost of sensors come down, and the ubiquitous use of GNSS-enabled smartphones or in-vehicle devices as sensors.

Good-quality data should meet four key attributes. The first is that of consistency – whereby sensors, regardless of its type, should give the same values for the same traffic conditions all the time. Having the same value to describe the same condition all the time is not sufficient. This is where the second attribute comes in, and this is accuracy. This means that the data must be correct in describing the traffic condition within its tolerance of accuracy. The key performance indicator for these two attributes can be indicated as, for example measured travel speeds must be within plus–minus 2 km/h of the actual value, 98% of the time. The third attribute is that of reliability and this is to ensure that data is available both spatially and temporally. For example, travel speed data for the road network should be available at enough locations on the roads and be available round the clock. If the data is not spatially comprehensive, it is often difficult to say that the eventual computed value is representative of the actual ground condition. Likewise, the computed information will be less useful if there are frequent periods of time that such data is not available. The fourth attribute is that the data should be exchangeable, primarily because data can come from many sources and sensor type. In the integrated i-transport platform that Singapore has, there is a defined standard for the various data in terms of its type and format, so that the data from one source can be used together with those from other sources and can also be used by many applications for its various functions.

Given the huge amount of data that is captured from a road network like Singapore's, it needs to be analysed and converted into meaningful information for it to be useful. Hence, like data, the quality of information is guided by three attributes. The first attribute is that of relevance. Information should be relevant to the traveller, for example a motorist on the western part of the city will not find information on congestion on the eastern part of the city very useful. The second attribute is that of timeliness. For example, information on traffic incidents on a road is useful to the traveller only if this is received early enough for this road-user to make detours to alternative route or transport mode – it is less useful if this same road-user is already trapped in the congested traffic arising from that incident. The third attribute is that of simplicity – information should be structured to be simple enough to be understood and read by the road-user within the short time available as there would be many other cognitive actions needed to safely drive the vehicle.

To sum up, the four key attributes for good-quality data are consistency, accuracy, reliability and exchangeable. That for good-quality information are relevance, timeliness and simplicity. This is a good practice that is adopted for the development of travel information from ITS in Singapore.

3.7.2   Travel information available from ITS in Singapore

With the implementation of EMAS on Singapore's expressways, speeds at various points became available from detector cameras installed at about 500-m intervals. These spot speeds were used to compute average travel times from one point of the expressway network to another, in almost real time. It became obvious that this information would be useful for motorists before they enter the expressway, as motorists would have the option of not entering the network if the travel time to their destination is unfavourable. Hence, EMAS includes electronic signboards before the entry into the expressway, displaying the travel times to several destinations (see Figure 3.6).

Initially the information displayed was in monotone and the occasional motorist would not be able to, from a number displayed as the travel time in minutes, distinguish between good and bad traffic conditions. This was because this occasional motorist would not know what is the normal travel time to reach the intended destination. This was subsequently improved when these signboards were upgraded to have the travel time numerals displayed in either green, amber or red, with each corresponding to normal, heavy or jammed traffic conditions.

When TrafficScan was subsequently implemented, the travel speeds provided by participating GNSS-enabled taxis were fused with EMAS data (where available) to provide average travel speeds on major roads and expressways. With this richer data set, traffic conditions on major roads and expressway could be displayed on maps, with colour-coded segments representing normal, heavy and jammed traffic conditions on the road network. This map-based information became available to the motoring public through the internet, and thereafter on smartphones. This was the initial use of

Figure 3.6   Roadside electronic signboard showing expected travel times to various destinations

vehicles as probes to provide traffic conditions on the roads. Subsequently, the ubiquitous smartphones became probes as well and the road network's traffic conditions computed therein became available on these same smartphones themselves.

The almost real-time availability of traffic conditions on the road network is not lost to those providing navigation services to motorists. With changing traffic conditions available to these navigation applications, they are able to give an alternative route to motorists arising from incidents on their earlier advised route to their destination. How the advised motorist behave in terms of their route choice thereafter is less clear. If motorists are fully compliant on the advice to change their route, it is probable that this alternative route will not be the best route after a short while. Hence, it is likely that some motorists will game the situation and stick onto the originally planned route, on the basis that with others diverting, this original planned route will be a better choice.

Using the same data analytics techniques, travel times of buses can also be determined and it follows that the expected arrival time of various bus services at the bus stops can be provided to its commuters. The information would show the expected arrival time of the approaching bus and the next one for each bus service. Initially, this information was displayed at electronic panels located at popular bus stops but again, as with information for drivers of private vehicles, the information subsequently became available on smartphones. Having the expected bus arrival times on smartphones has the added advantage of the bus commuters being able to check on this piece of information well before reaching the bus stop, and on whether slow walking is sufficient or a quick short run is necessary.

In Singapore, commuters tap their transport smart card onto readers on boarding and alighting the buses. This is necessary as bus fares are distance-based, and the tap-in tap-out process allows the correct fare to be deducted. However, there is an added feature possible with this process – the number of passengers on board the bus can be determined, and hence information on crowdedness of the bus can be provided, together with the expected bus arrival times, to waiting bus commuters. The commuters can then decide if they should board the approaching bus or the next bus based on this extra piece of information.

As more sensors became available to monitor the traffic condition on the road network, it becomes possible to use the massive amount of data to predict what the traffic condition will be like in the near-term (in the next 10–30 min) as a result of an incident happening. This is, of course, contingent on knowing how motorists will behave for various scenarios. The benefit from being able to know what the traffic condition will be like in the near-term is that transport and traffic operators can pre-empt and apply mitigating measures well before the road network gets clogged up as a result of that incident. All these are possible with the advent of faster computers and the application of artificial intelligence algorithms, and these approaches are being explored presently.

3.8   Connected and autonomous vehicles

Public transport will have to remain the main mode of transport in Singapore and given the manpower constraints faced by an ageing workforce, the operations of the public transport have to rely less on human-intensive activities. Unlike the MRT system, bus operations presently rely on human drivers to move them. This limits the ability to put in more buses on the roads as travel demand increases over time, since additional buses need additional drivers.

The advances in autonomous vehicles that do away with the need for drivers are therefore of significant relevance to Singapore's transport strategy. The focus will have to be on autonomous buses or shuttles, and not on private cars. It will have a key role in replacing the current human-driven high-capacity feeder buses that bring commuters within residential towns to the MRT stations or bus interchanges. For parts of residential towns that are lower in density, autonomous shuttles with lower carrying capacity may well be useful to connect commuters to the public transport nodes. Plans are in progress to allow pilot deployment of autonomous buses and shuttles in three new communities from 2022 – the residential towns of Punggol and Tengah and the Jurong Innovation District.

To ensure that this new development is given its due attention and recognition, there is a high-level committee headed by the Permanent Secretary at the Ministry of Transport to steer this. Comprising representatives from a diverse range of Ministries and Statutory Boards, it also included experts from the industry and the academia.

Already, a road network in an R&D hub in Singapore (called one-north) has seen various autonomous vehicle developers doing trials there with their vehicles in a mixed traffic environment. These autonomous vehicles have to, however, undergo 'milestone' tests before they are allowed to be on the roads in this one-north area. Additional 'milestone' tests are also needed for the autonomous vehicles (and the software that drives these vehicles) before they are allowed to venture to other mixed traffic routes outside one-north area. These more stringent tests are needed because outside of one-north, they will come onto residential roads as well as near schools and other more sensitive developments.

To facilitate the 'milestone' and other tests, the Government has developed an off-road 2-ha test facility called CETRAN. Other tests conducted there are on the effectiveness of communication equipment to connect between infrastructure and vehicles. While it is necessary for autonomous vehicles to operate safely at all times, this often comes with trade-offs on efficiency, and the usefulness of connecting these vehicles to infrastructure (e.g. to provide status of traffic signals, or the remaining time to red) is to bring back some of these efficiencies while retaining its high safety level.

Autonomous vehicles are envisaged not just for public transport but also to facilitate point-to-point mobility-on-demand services. This is in line with the concept of mobility-as-a-service (MaaS) where vehicles are for shared use and made available on demand by commuters using an integrated smartphone-based application. Several combinations of transport modes may be proposed with different travel experiences and price points for the subscriber to choose. The motivation for MaaS, from Singapore's perspective, is that it is not tied to the concept of vehicle ownership.

Logistics vehicles such as container trucks and delivery vehicles are also possible beneficiaries of autonomous vehicles, whereby its mode of operations is no longer restricted to the availability of drivers. Likewise, autonomous vehicles are also envisaged to be used for various utility functions such as road sweeping and cleaning.

3.9 Concluding remarks

Together with the public-transport-focused land transport strategy for Singapore, ITS has been a useful and effective technology to keep traffic on the road network flowing. The respectable traffic situation that Singapore enjoys today is evident from various third-party reports such as the INRIX 2017 Global Traffic Scorecard [2] and the McKinsey & Company's report on urban transportation systems of 24 global cities [3]. The former ranked Singapore 181st out of 200 cities for congestion, with less than 10 h spent in congestion during peak periods in 2017. McKinsey & Company's report has Singapore as a top city in its overarching urban mobility ranking. Nevertheless, there is still much that can be done to improve the travel experiences of commuters in Singapore.

References

[1] Land Transport Authority Singapore, Land Transport Master Plan 2013; 2013.

[2] INRIX, Global Traffic Scorecard; 2017. http://www2.inrix.com/us-traffic-hotspot-study-2017; 2017 [Accessed 12 Oct 2018].

[3] Knupfer, S.M.; Pokotilo, V.; and Woetzel, J., *Elements of Success: Urban Transportation Systems of 24 Global Cities*, McKinsey & Company, 2018.

[4] Chin K.K., *The GLIDE System – Singapore's Urban Traffic Control System*, Transport Reviews, 13:4, 295–305; 1993. DOI: 10.1080/01441649308716854.

[5] Chin, K.K., *Road Pricing – Singapore's 30 Years of Experience*, CESinfo DICE Report, Journal for Institutional Comparisons, Munich, 3:3, 12–16; 2005.

[6]   Watson, P.; and Holland, E., *Relieving traffic congestion: the Singapore area license scheme.* Staff Working paper no. SWP 281, Washington, DC, The World Bank, 1978; http://documents.worldbank.org/curated/en/88318146 8759586286/Relieving-traffic-congestion-the-Singapore-area-license-scheme.

[7]   Menon, A.P.G; and Loh, N., Singapore's Road Pricing Journey – Lessons Learnt and Way Forward; Journeys Issue 14; November 2015, Land Transport Authority Singapore.

[8]   Land Transport Authority Singapore, Smart Mobility 2030 – ITS Strategic Plan for Singapore; 2014.

Part II

Traffic state sensing by roadside unit

Chapter 4

Traffic counting by stereo camera

Toshio Sato[1]*

4.1 Introduction

Computational stereo vision introduced by Marr [1] is one of the powerful methods to extract three-dimensional (3D) information using multiple images. By applying stereo vision, many kinds of traffic monitoring systems have been developed. In this chapter, we discuss configurations and characteristic of stereo-based traffic monitoring systems especially for the roadside unit.

Stereo vision basically measures depth from disparity between the left and right camera images based on triangulation. In the early 1980s, researches have started to develop computational stereo vision systems that measure depth automatically [2–4]. The automatic measurement is realized by finding correspondences between points of two images such as block-matching techniques. Correspondence between two images makes a disparity map and we can transform this into a depth map that represents distances between a camera and forward objects. Camera calibration [5] and transformation to the epipolar standard geometry are also developed to achieve practical stereo vision systems.

Implementation of stereo vision is computationally expensive especially for block-matching process. This subject induced many studies to build hardware-based stereo systems [6]. In the 2000s, processing power of PCs became higher enough to implement real-time stereo vision by software. Image processing libraries OpenCV include functions related to stereo vision [7].

We need other processes to build stereo vision systems that count moving objects with precision. For instance, video-rate object tracking using sequential depth maps is one of the key processes to realize practical stereo vision systems.

Since stereo vision is robust against illumination variations compared with simple pattern matching, there are many applications for outfield uses such as driving assistance for vehicles and object detection for robots. Traffic state monitoring is one of the prevalent applications of stereo vision that detects and counts vehicles or humans. Though the basic idea of stereo-based vehicle detection is reported in 1978 [8], real-time processing systems have been developed in the early 1990s [9–11]. Among applications of vehicle detection, drive assistance using onboard stereo cameras is

[1]Toshiba Corporation, Power and Industrial Systems Research and Development Center, Kawasaki, Japan
*Current Position: Waseda University, Tokyo, Japan

the major research area and there are few roadside stereo systems for traffic monitoring. On the other hand, there are many researches on stereo-based people counters using fixed-point camera setups [12]. Fixed-point observation has advantages for accurate counting of objects and it is considered that vehicle counting using stereo vision for the roadside equipment also has advantageous conditions for accuracy.

This chapter conducts approaches to build an accurate traffic counting system using stereo vision. In Section 4.2, general procedures of stereo vision system for traffic monitoring is explained. Then an example of the accurate traffic counting system is described as one of the applications of stereo vision in Section 4.3. Finally, comparison of the traffic counters is discussed based on accuracy and its costs.

4.2 General procedure traffic counting using stereo vision

4.2.1 Stereo cameras

The stereo vision system always uses two cameras as shown in Figure 4.1, to obtain two differing views. In Figure 4.1, two cameras cam_R and cam_L are mounted separating in a baseline length b.

The angle of view of the camera and the baseline length b should be carefully designed to obtain the sufficient field of view and resolution to detect objects. As shown in Figure 4.2, a detection area should be covered by common fields of both cameras. The minimum distance of the common field of view Z_0 is calculated by the camera angle a_{cam} and the baseline length b:

$$Z_0 = \frac{b/2}{\tan(a_{cam}/2)}$$

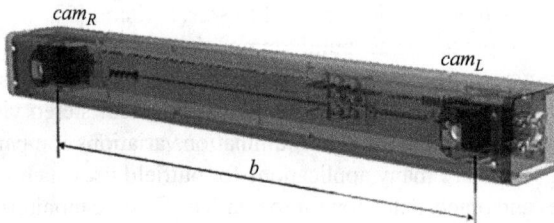

Figure 4.1   A typical composition of the stereo camera

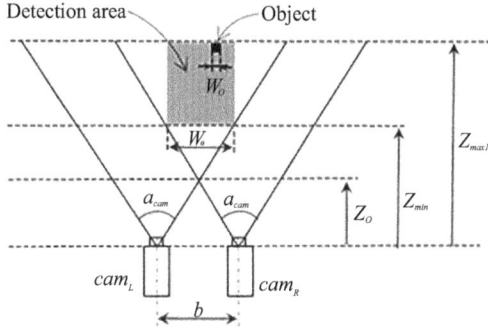

Figure 4.2   Installation and a field of view of the stereo camera (top view)

The minimum depth Z_{min} that contains the width of detection area w_a is calculated as follows:

$$Z_{min} = \frac{Z_0 + w_a/2}{\tan(a_{cam}/2)}$$

$$= \frac{w_a + b}{2\tan(a_{cam}/2)}$$

The maximum depth Z_{max} is determined by two conditions. The first condition is image resolution to capture an object. When the number of pixels in the horizontal direction of the camera is N_s, the width of an object w_o should be larger than the size for 1 pixel of the image (i.e., resolution) at the depth Z_{max1} (see Figure 4.2):

$$w_o \geq \frac{2Z_{max1}\tan(a_{cam}/2)}{N_s}$$

Hence, Z_{max1} defined by the first condition is calculated as follows:

$$Z_{max\ 1} = \frac{w_o N_s}{2\tan(a_{cam}/2)}$$

The second condition relates to the minimum disparity at the maximum depth Z_{max2}. The depth Z is calculated by disparity d based on a focal length f and a baseline length b in general stereo vision principles [7].

$$Z = \frac{fb}{d} \qquad (4.1)$$

The disparity d is displacement of corresponding points in image planes between X_L and X_R as shown in Figure 4.3. The disparity and depth are explained in detail in the next section. The minimum disparity d_{min} depends on the cell size of the sensor that is calculated by the width of the sensor w_s and the number of pixels N_s:

$$d_{min} = \frac{w_s}{N_s} \text{ (mm/pixel)}$$

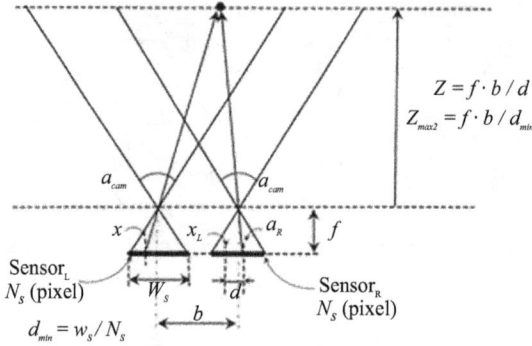

Figure 4.3   Installation and minimum disparity of the stereo camera (top view)

Therefore, the Z_{max2} defined by the second condition is calculated as follows:

$$Z_{max\,2} = \frac{fb}{d_{min}} = \frac{fbN_s}{w_s}$$

Based on the configuration of Figure 4.3, the angle a_{cam} is obtained by the width of the sensor w_s and the focal length f:

$$a_{cam} = 2\arctan\left(\frac{w_s}{2f}\right)$$

According to the equations described previously, under the conditions of $b = 400$ mm, $w_s = 4.8$ mm, $N_s = 640$ pixel, $w_a = 3{,}000$ mm and $w_o = 500$ mm, Z_{min}, Z_{max1} and Z_{max2} are estimated as 3.2, 85 and 240 m, respectively. Though $Z_{max2} = 240$ m is the relatively longer value, the resolution of depth should be considered for practical uses. Figure 4.4 shows a relationship between disparity and depth for the previous conditions. In this figure, disparity dx are described as the number of pixels that are transformed by $dx = d/d_{min}$. At maximum depth $Z = 240$ m, the change of one disparity causes depth errors of over 100 m. According to Figure 4.4, the estimation value $Z_{max2} = 240$ m should be revised as approximately 50 m within depth errors of 10 m. Totally, the designed stereo camera will cover 3 m in width and depth from 3.2 to 50 m under the previous conditions.

There are other important points to design a stereo camera. Two cameras should be installed in parallel fixing on a rigid basement and two sensors should be synchronized by using external triggers to capture right and left images simultaneously.

4.2.2   Calibration of camera images

The first process that the stereo vision system should execute is calibration of camera images. Camera images are always distorted, including radial distortion and tangential distortion. These distortions occur by a lens itself and the alignment between a lens and a sensor. Figure 4.5 shows examples of radial distortion mainly caused by the lens. An image of the object is distorted as "fish-eye" and "barrel" distortions through the camera module.

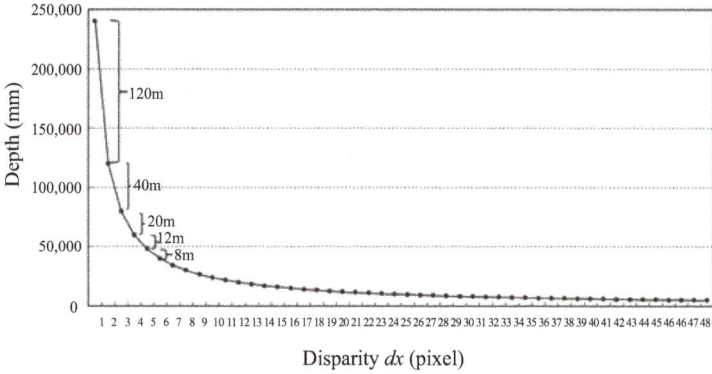

Figure 4.4   Relationship of disparity and depth

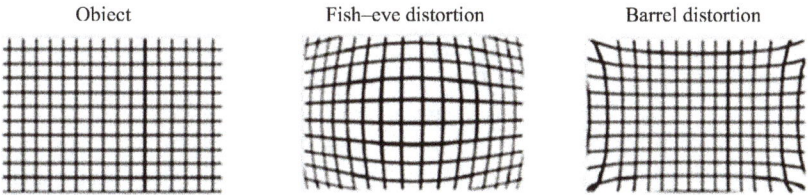

Figure 4.5   Examples of lens distortion

Figure 4.6   An example of calibration pattern (a chessboard)

The distortion is compensated by a transformation matrix that is calculated using relationship of feature points in known calibration patterns and ones of captured images. Figure 4.6 illustrates a typical calibration pattern called "a chessboard." Calibration is executed by using multiple frames of the calibration grid with various positions and angles. The automatic calibration method proposed by Tsai [5] and an advanced method proposed by Zhang [13] are now available in OpenCV libraries [7].

4.2.3   Image rectification

Another transformation for camera images is "rectification" [7] that transforms two images in the same plane and vertically aligned. Two images are transformed into the epipolar standard geometry as shown in Figure 4.7. In the epipolar standard geometry, both projected image planes are reconstructed to lie on the same plane and a point of the object $P(X,Y,Z)$ is projected to $P_L(X_L, Y_L)$ and $P_R(X_R, Y_R)$ on the image plane where the Y-coordinate value of an object is the same ($Y_L = Y_R$). Image rectification simplifies the geometry of the system and greatly reduces calculation costs. For instance, corresponding points lie on the same vertical coordinate of the left and right images and the correspondence search can be limited in one-dimensional direction.

4.2.4   Block matching to produce a depth map

To find correspondences of feature points of the object in two images, block matching in a local area (such as a 3×3-pixel area) is applied. In the epipolar standard geometry, the block matching searches the highest similarity or the lowest difference along the horizontal coordinate Le (the epipolar line) in Figure 4.8.

A local area centered at an arbitrary pixel of the left image $p_L(x,y)$ is selected as a template $Tm(i,j)$ and the template $Tm(i,j)$ is searched for the best matching position in the right image $p_R(x,y)$. The $p_L(x,y)$ and $p_R(x,y)$ are descriptions of pixels on the digital images for the virtual $P_L(X_L, Y_L)$ and $P_R(X_R, Y_R)$, where the length of x and y is not measured based on distances but the number of pixels. The best matching

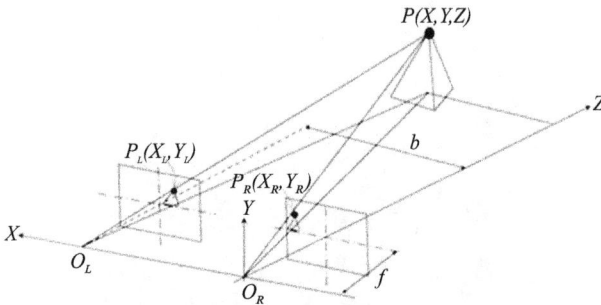

Figure 4.7   Epipolar standard geometry

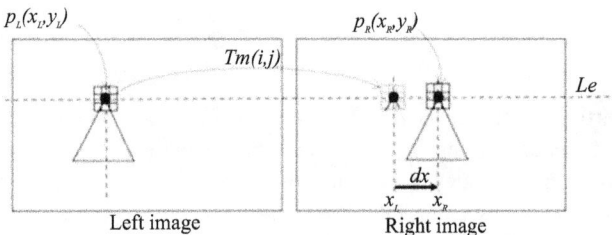

Figure 4.8   Block matching between stereo images

position is defined by a metrics such as sum of absolute differences (SAD) and sum of squared differences (SSD). The SAD and the SSD are standard approaches because of their simple calculation procedures as shown next:

$$S_{SAD}(dx, dy) = \sum_{i} \sum_{j} |p_R(dx + i, dy + j) - Tm(i,j)|$$

$$S_{SSD}(dx, dy) = \sum_{i} \sum_{j} (p_R(dx + i, dy + j) - Tm(i,j))^2$$

where i and j are variables that change in width and height of the template $Tm(i,j)$, respectively. Variables dx and dy are shift values in a search range. In the epipolar standard geometry, the search range is limited on the epipolar line Le, which means dy can be fixed to "0." The minimum value of SAD or SSD is selected as the best match point and the shift value dx is calculated by $dx = x_R - x_L$ as the equivalent disparity counted by pixels. A disparity map is produced calculating dx for each pixel of the image.

Based on a cell size of the sensor w_c, a relationship between dx and d is described as follows:

$$d = dx \times w_c$$

where w_c is calculated by using the width of a sensor w_s and the number of pixels of the sensor Ns:

$$w_c = \frac{w_s}{N_s}$$

According to (4.1) in Section 4.2.1, the disparity d corresponds to depth Z from the stereo camera that has the properties of the focal length f and the baseline length b. A depth map is produced from the disparity map based on (4.1). The values in this disparity map are inversely proportional to the scene depth at the corresponding pixel location. In many cases, the disparity maps are used for object detection without transformation to a depth map.

Input image for the block matching is often standardized in intensity values using histogram equalization, for instance, to handle illumination changes. Edge enhancement such as the Sobel filter is also applied to detect feature points with stability. The postfiltering process for the results of the block matching is applied to eliminate bad correspondence matches comparing the metric and a threshold.

4.2.5 Object detection

As the next procedure, objects such as vehicles and pedestrians are detected using a depth map or a disparity map. In many cases, the depth of the road surface appears in a plane distribution as shown Z_1–Z_6 in Figure 4.9. By contrast, obstacles or objects have different properties along the vertical coordinates in the depth map. The depth of an object Z_{obj} is larger than one of the background area Z_{back} corresponding to the road surface. Based on this difference in depth, objects are detected to examine the depth map along the horizontal coordinates L_h. Some statistical approaches, including calculation of the average in a local region, are applied to select appropriate depth and improve robustness of detection as shown in Figure 4.9(b).

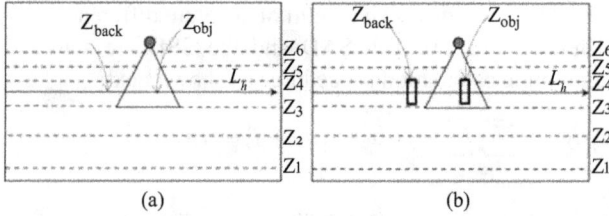

Figure 4.9   *Object detection based on difference of depth. (a) Two depth values along L_h. (b) Using a local region (a rectangle window)*

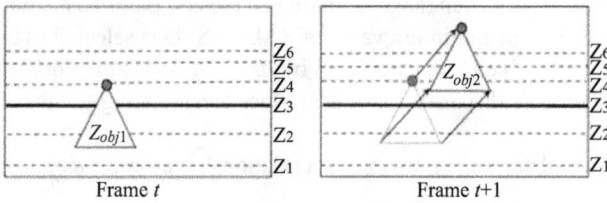

Figure 4.10   *Object tracking and counting across a virtual line D_3*

In the case of using a disparity map instead of the depth map, the basic idea to detect objects is the same.

4.2.6   Object tracking and counting

A moving object appears in multiple depth maps for the input video data changing its depth value. To count a moving vehicle correctly, object tracking is necessary through the multiple frames. Figure 4.10 illustrates a simple method to count a vehicle. In Figure 4.10, an example of a moving object is shown that changes its depth from Z_{obj1} to Z_{obj2} in the frames t to $t + 1$. Determined a virtual line on the depth D_3, a target that moves forward is counted if $Z_{obj1} < Z_3$ and changed to $Z_{obj2} > Z_3$. Note that this relationship will be reversed if the system uses a disparity map instead of the depth map.

In actual traffic monitoring systems, much complicated rules are designed to handle traffic congestion or backward movement of vehicles or humans.

4.2.7   Installation of stereo camera

The setup of the stereo camera unit is important to satisfy various purposes of traffic monitoring. There are some ways to determine the field of view of the stereo camera as shown in Figure 4.11. Figure 4.11(a) is a typical field of view for the onboard camera of vehicles. As described in Section 4.2.5, object detection is performed based on difference of depth between vehicles and the road surface. By contrast, it is difficult to detect a second vehicles occluded by the closest one. Figure 4.11(b) is one of the examples for the roadside systems. Its field of view is the same as conventional roadside video surveillance systems

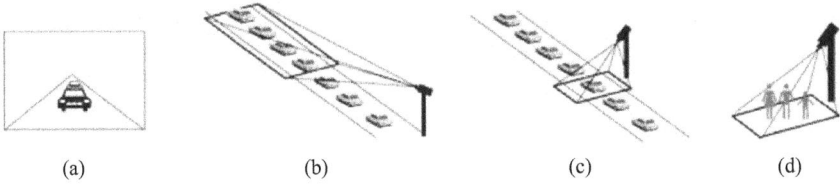

(a) (b) (c) (d)

Figure 4.11   Installation of stereo cameras for traffic monitoring. (a) On-board camera [10], (b) Roadside video surveillance [11], (c) Roadside vehicle counting [13], (d) People counting [12]

that do not use stereo cameras but monocular cameras. In this case, the system can detect vehicles in a wider area; however, the accuracy of depth measurement tends to be inferior. Installation of Figure 4.11(b) has longer depth area and less disparity resolution will be obtained by the stereo camera. Installation of Figure 4.11(b) includes another problem of difficulties to acquire the depth of the road surface.

Figure 4.11(c) illustrates another installation of the roadside stereo camera. Using this setting, accurate vehicle counting is expected to use depth of the road surface [14], though the field of view is smaller than Figure 4.11(b). People counters [12] use the same strategy to capture objects and the surface in the same field of view as shown Figure 4.11(d).

In the next section, an example of a vehicle counting system by using installation of Figure 4.11(c) is explained in detail.

4.3   Accurate vehicle counting using roadside stereo camera [14]

In this section, an application of stereo vision for vehicle counting is explained. The basic algorithm is the same as methods described in Section 4.2.

Vehicle detection is an essential function for toll collection systems that counts the number of vehicles and controls components such as toll gates or wireless charging for each transit. Inductive loop sensors, magnetic sensors, arrays of transmission sensors or laser scanners are used for vehicle detection [15]; however, there are problems for equipment, installation and maintenance costs. Inductive loop sensors and magnetic sensors are needed to be buried under road surfaces to increase construction costs. Though laser scanners can be installed simply at the top or the side of a road, equipment cost is expensive and mean time between failure (MTBF) is relatively lower. Moreover, most sensors provide only one-dimensional scanning data that does not supply detailed information such as moving speed and directions. To detect moving directions, two sensors are needed that make much expensive costs.

Video image processing for vehicle detection is studied for simple low-cost applications. Forward vehicle detection using stereo vision is reported for safety or automatic driving [16,17]. In the most reports, the detection rates of stereo-vision-based detection are lower than 99% [17] that are insufficient for toll collection system.

To solve these problems, a high-accuracy stereo vision system that can be settled beside a road is introduced. An arrangement of stereo cameras and algorithms of vehicle detection is described to achieve higher performance.

4.3.1   System configuration

To achieve higher performance of vehicle detection, a specialized configuration of stereo cameras is designed. The circle of Figure 4.12 indicates a stereo camera module that is fixed at 3.0-m high with a mast built beside a pathway. Stereo cameras are arranged in the longitudinal direction, and scanning lines of each industrial television camera (ITV camera) are adjusted parallel with the longitudinal directions. Figure 4.13 shows an example image captured using the stereo camera module. Each image has 240-pixel (horizontal) and 640-pixel (vertical) resolutions, which covers 3.5-m width for a pathway. In Figure 4.13, width of image is designed to acquire several frames for a passing vehicle with 100-km/h speed using 15 frames per second.

Infrared LED lights are attached for operation at night and the cameras have sensitivities for infrared light.

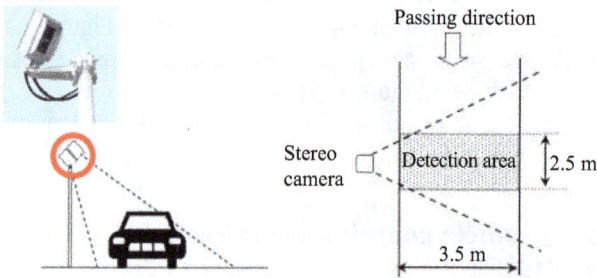

Figure 4.12   Installation of stereo camera module (left) and a top view of detection area on road surface (right)

Figure 4.13   Example of stereo image (left: lower camera, right: upper camera)

Algorithms of vehicle detection consist of depth measurement, vehicle detection and a traffic counter.

4.3.2 Depth measurement based on binocular stereo vision

Binocular camera images are captured for the same scene to obtain 3D image, as shown in Figure 4.14. P_1 and P_2 are points on image planes for scene points P on an object, and the depth of scene points P can be calculated to strike disparities between P_1 and P_2 on the image plane. Two cameras using equivalent focal length are set up and their optical axes are adjusted as parallel to each other.

Since images captured by the binocular camera are not well aligned, rectification for the input images is applied. Homographies are generated to transform the images in advance. These homographies can be applied to each frame thorough the whole images to rectify them correctly.

As shown in Figures 4.15 and 4.16, disparities between the transformed images are corresponded with the depth of the distance from the camera module.

Block-matching process between two images is applied to calculate disparities.

Edge detection using Sobel operator is applied to the both images and the matching process select a pixel of upper image based on SAD matching values. Sobel edge detection is used to extract edges of vehicle. Moreover, various constrains are applied to remove the incorrect matching candidates.

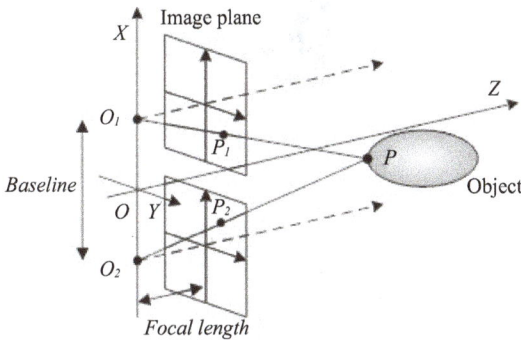

Figure 4.14   Pairs of camera configuration

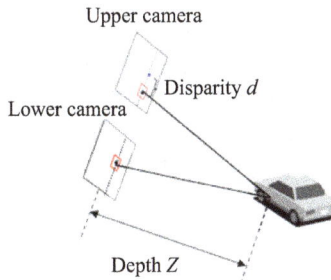

Figure 4.15   Distance measured by disparities of stereo vision

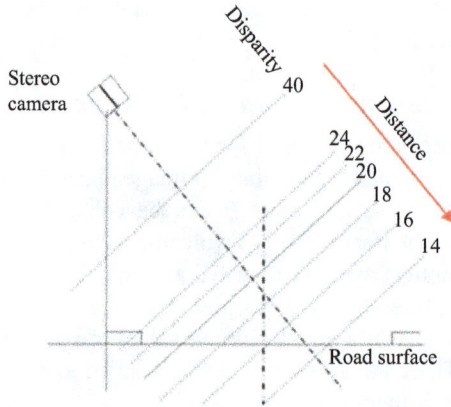

Figure 4.16   Relationship between the disparity value and distance

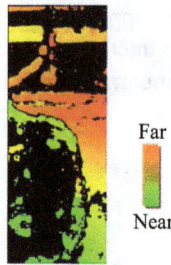

Figure 4.17   Example of depth map

Depth information is calculated using disparities, as shown in Figure 4.17. Figure 4.17 includes depth information, especially contour of vehicle and road.

4.3.3   Vehicle detection

Vehicle detection confirms the existence of a vehicle using the depth map. Since there are differences in depth between the vehicle and the road, candidates of the vehicles using subtraction of input depth data and background road data are extracted. The background data is made using depth values of the road (with no vehicle passing).

Three detection areas are designed to determine existences of a vehicle as shown in blue rectangular frames of Figure 4.18. Subtraction of depth data is evaluated in three detection areas. Figure 4.18 shows an example image of two vehicles in conjunction and results of detection indicate existences of the vehicle in "Area 1" and "Area 3."

4.3.4   Traffic counter

The traffic counter classifies states of detection area and verifies transition of the states of detection to achieve higher counting accuracy.

Five states are assigned to results of the vehicle detection in the three areas, as shown in Table 4.1. In the case of Figure 4.18, the state is assigned to "S4" that represents of two vehicles in conjunction.

Transitions of the states are verified to count vehicles. Basic rules of the transition for forward passing are described in Figure 4.19. In the most cases of forward passing, the state transit in "S0, S1, S2, S3, S0." In the case of Figure 4.18, the state will transit in "S0, S1, S2, S4."

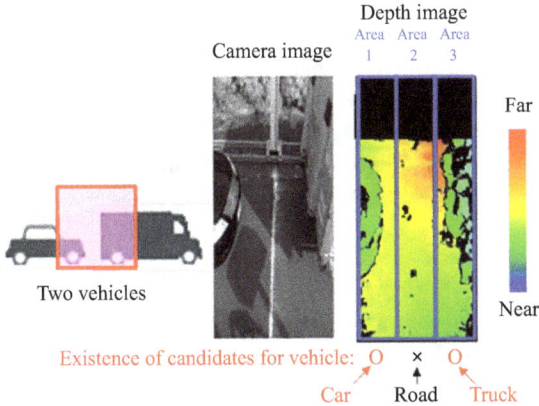

Figure 4.18   Subtract depth image and detection area

Table 4.1   States of vehicle detection

States		Existence of candidates for vehicle		
		Area 1	**Area 2**	**Area 3**
S0	No vehicle	×	×	×
S1	Entry	○	×	×
S2	Stay	×	○	×
		○	○	×
		×	○	○
		○	○	○
S3	Exit	×	×	○
S4	Two vehicles	○	×	○

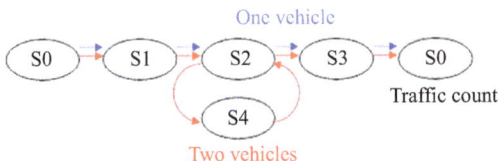

Figure 4.19   State transition diagram of vehicle detection

The algorithms are implemented as software on a PC (CPU: Intel® Core™ i5 2.4 GHz) with a video capture boards for ITV cameras. The vehicle detection is executable in 15 frames per second using this system.

4.3.5   Results

The system is designed to detect minimum 500-mm object; however, there are many illegal patterns for traffic counting. A tractor that is towed by a towbar of 60 mm in diameter should be counted as one vehicle. The minimum distance between sequential two vehicles to count two vehicles is defined not to merge two vehicles into one. Counting area is of 230–1,600 mm.

The system is confirmed its basic function using example in closed test environment. Examples of Figure 4.20 are simulated examples using people and bicycles.

The system is tested totally through the minimum size of a vehicle, illumination change and noises such as rain and snow.

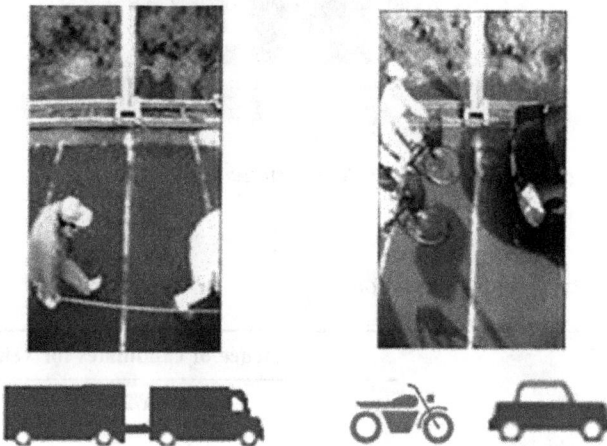

Figure 4.20   Simulated examples to confirm difficult vehicle patterns (left: tractor towed by a towbar, right: close vehicles in a congested situation)

Table 4.2   Experimental results

Date	Total vehicles	Hit numbers
2011/8/5 05:54–19:52	4884	4884
2011/8/22 11:40–19:09	2285	2286
2011/9/1 16:34–19:52	932	932
2011/9/21 17:02–19:53	820	820
2011/9/22 17:00–19:54	904	904
2011/9/23 17:02–18:58	441	441
Total	10266	10267
Detection rate		99.99%

Table 4.3   Comparison of traffic counter

	Inductive loop detector	Ultrasonic sensor	Microwave vehicle sensor	Infrared sensor	LIDAR	Single camera	Onboard stereo camera	Roadside stereo camera
Typical count accuracy	97% [18]	98% [18]	95%–98% [18]	99% [18]	99.5% [19]	95% [18]	99% [17]	99.99% [14]
Equipment cost	$$	$	$	$	$$$	$	$$	$$
Installation cost	High (underground)	Low (roadside)	Low (roadside)	Low (roadside)	Low (roadside)	Low (roadside)	Low (onboard)	Low (roadside)

The system is installed at actual toll correction station and evaluated using 10,266 traffics, including early morning through night data. Traffic data consists of various kinds of vehicles, including bikes through buses and trucks. Table 4.2 shows results that achieve higher detection rate, 99.99%, compared with conventional stereo vision approach. An error is caused by overdetection that can be adjusted parameters of algorithms.

The result indicates that the stereo-based traffic counting achieves highly accurate detection rate. Stereo cameras and algorithms are designed for toll collection system and experimental results indicate 99.99% detection rate. It is considered that this approach realizes low equipment and installation costs for vehicle detection of toll collection systems.

4.4 Summary

In this chapter, the general procedures and an application of stereo vision for the roadside vehicle counting are discussed.

At the moment, stereo vision is prevalent for onboard object detection utilized to driving assistance. However, the roadside stereo vision system has possibilities to achieve accurate traffic counting performance up to 99.99% detection rate. Typical precision rates of the most traffic counters, including inductive loop detector, microwave vehicle sensor and ultrasonic sensor, are 99% or less [18].

It is considered that pneumatic road tubes, piezoelectric sensors and optics fibers have higher detection rates; however, these sensors can detect existence of wheels' pressure and have limitation to count the number of various types of vehicles with high accuracy. Furthermore, there involve problems of installation costs to set underground. The inductive loop sensors also have this problem, though they count vehicle body with accuracy over 97% [18].

Table 4.3 explains comparison results of the traffic counter. The roadside stereo vision represents higher precision rates compared with onboard vehicle detection of driving assistance, because of its rigid constraint of standard distances to the road surface and moving direction of vehicles. LIDARs can also measure depth and have possibilities to realize accurate vehicle counting [19]; however, the equipment cost is much higher than the roadside stereo vision.

According to comparison in Table 4.3, the roadside stereo vision system has advantages that achieve accurate vehicle counting performance with reasonable equipment and installation costs.

References

[1] Marr D. and Poggio T. 'A computational theory of human stereo vision'. *Proceedings of the Royal Society of London. Series B, Biological Sciences*; 1979; 204(1156). pp. 301–328.

[2] Grimson W.E.L. 'A computer implementation of a theory of human stereo vision'. *Philosophical Transactions of the Royal Society of London. Series B, Biological Sciences*; 1981; 292(1058). pp. 217–253.

[3] Baker H.H. and Binford T.O. 'Depth from edge and intensity based stereo'. *Proceedings of the 7th International Joint Conference on Artificial Intelligence – Volume 2*; Vancouver, Canada, August 1981, pp. 631–636.

[4] Lucas B.D. and Kanade T. 'An iterative image registration technique with an application to stereo vision'. *Proceedings of the 7th International Joint Conference on Artificial Intelligence – Volume 2*; Vancouver, Canada, August 1981, pp. 674–679.

[5] Tsai R.Y. 'A versatile camera calibration technique for high-accuracy 3D machine vision metrology using off-the-shelf TV cameras and lenses'. *IEEE Journal of Robotics and Automation*; 1987; 3(4). pp. 323–344.

[6] Kanade T., Yoshida A., Oda K., Kano H. and Tanaka M., 'A stereo machine for video-rate dense depth mapping and its new applications'. *Proceedings of 15th Computer Vision and Pattern Recognition Conference*; San Francisco, USA, June 1996, pp. 196–202.

[7] Bradski G. and Kaehler A. *Learning OpenCV: Computer Vision With the OpenCV Library*; O'Reilly Media, Sebastopol, USA; 2008. Chapter 11–12.

[8] Gennery D.B. 'A stereo vision system for an autonomous vehicle'. *Proceedings of the 5th International Joint Conference on Artificial Intelligence – Volume 2*; Cambridge, USA, August 1977, pp. 576–582.

[9] Luong Q.-T., Weber J., Koller D., and Malik J. 'An integrated stereo-based approach to automatic vehicle guidance'. *Proceedings of IEEE International Conference on Computer Vision*; Cambridge, USA, June 1995, pp. 52–57.

[10] Bohrer S., Zielke T., and Freiburg V. 'An integrated obstacle detection framework for intelligent cruise control on motorways'. *Proceedings of the Intelligent Vehicles'95. Symposium*; Detroit, USA, September 1995, pp. 276–281.

[11] Malik J., Weber J., Luong Q.-T., and Koller D. 'Smart cars and smart roads'. *Proceedings of the 6th British Conference on Machine vision (Vol. 2)*; Birmingham, UK, July 1995, pp. 367–382.

[12] Terada K., Yoshida D., Oe S., and Yamaguchi J. 'A method of counting the passing people by using the stereo images'. *Proceedings of 1999 International Conference on Image Processing Vol. 2*; Kobe, Japan, October 1999, pp. 338–342.

[13] Zhang Z. 'A flexible new technique for camera calibration'. *IEEE Transactions on Pattern Analysis and Machine Intelligence*; 2000; 22(11). pp. 1330–1334.

[14] Takahashi Y., Aoki Y., Hashiya S., Kusano. A., Sueki N, and Sato T. 'Accurate vehicle detection using stereo vision for toll collection systems'. *Proceedings of 19th ITS World Congress*; Vienna, Austria, October 2012, AP-00280.

[15] Klein L., Mills M., and Gibson D. '*Traffic Detector Handbook: Third Edition – Volume I*'. FHWA-HRT-06-108, Federal Highway Administration, USDOT USA; 2006.

[16] Bertozzi M, Broggi A., Fascioli A, and Nichele S. 'Stereo vision-based vehicle detection'. *Proceedings of the IEEE Intelligent Vehicles Symposium 2000*; Dearborn, USA, October 2000, pp. 39–44.

[17]   Kowsari T., Beauchemin S., and Cho J. 'Real-time vehicle detection and tracking using stereo vision and multi-view AdaBoost'. *Proceedings of 14th International IEEE Conference on Intelligent Transportation Systems*; Washington, DC, USA, October 2011, pp. 1255–1260.

[18]   Zhan F., Wan X., Cheng Y., and Ran B. 'Methods for multi-type sensor allocations along a freeway corridor'. *IEEE Intelligent Transport Systems Magazine*; 2018; 10(2). pp. 134–149.

[19]   Sato T., Aoki Y., and Takebayashi Y. 'Vehicle axle counting using two LIDARs for toll collection systems'. *Proceedings of 21st ITS World Congress*; Detroit, USA, September 2014, 12249.

Chapter 5

Vehicle detection at intersections by LIDAR system

Hikaru Ishikawa[1], Yoshihisa Yamauchi[1] and Kentaro Mizouchi[1]

5.1 Introduction

5.1.1 New trend

Intersection monitoring using LIDAR (light detection and ranging) system has been performed to a limited extent until recent years because of the higher cost of the LIDAR hardware compared with cameras and the reason that the higher cost did not promote the development of point cloud technology.

However, opinions on LIDAR system are changing. Autonomous vehicles have been adopting LIDAR system as key sensors, and the robust and reliable measurements from the LIDAR system are highly evaluated. It is boosting the development of point cloud technology. There is also expectation that a high-volume market for the autonomous vehicles will lower practical LIDAR price in the future. Vehicle detection at intersection by the LIDAR system is beginning to appear as a new trend.

5.1.2 Target applications

The LIDAR system measures the three-dimensional (3D) position and shape of vehicles and the geographical situation of intersection. By analyzing the information, it is possible to identify vehicle types and sizes, and the traffic characteristics of vehicle movements.

Reconstructed traffic characteristics are expected to be used for E-Tolling, traffic flow measurement and optimization, intersection safety, traffic law enforcement, and intersection environment management. Figure 5.1 shows how the data measured by the LIDAR system are used in each target application.

5.1.3 Basic principle of LIDAR system

LIDAR system illuminates light in a specific direction and measures the time it is reflected back from the object (time of flight=TOF). Since the propagation speed of light in air (in vacuum) is fixed, the reciprocating distance to the

[1]Intelligent Information Management Headquarters, IHI Corporation, Tokyo, Japan

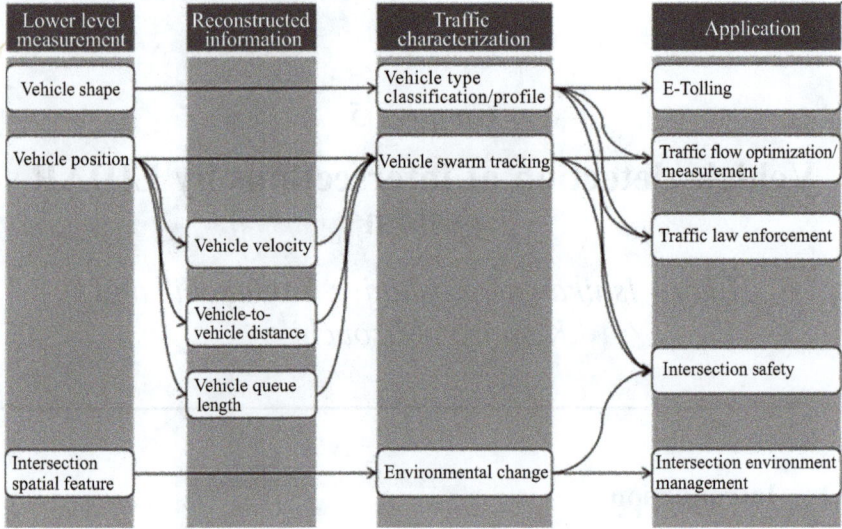

Figure 5.1   Target applications and measured data

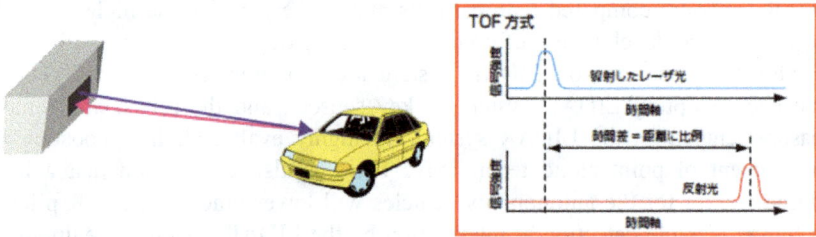

Figure 5.2   Time-of-flight measurement

object can be obtained by dividing the elapsed time by the light velocity as shown in Figure 5.2.

By repeating the measurements while changing the projecting direction sequentially (Figure 5.3), information of object existence and the shape can be obtained.

5.1.4   Types of LIDAR system

Based on the laser sweep method, LIDAR system can be classified into two types: 2D and 3D. The 3D one is further classified as short- and long-range types as shown in Figure 5.4.

In the 2D type, the laser is swept in one direction and line scanning is performed by the light section method. In the 3D method, the measurement is performed by sweeping the laser in two axial directions or sweeping in one axial direction and projecting a plurality of lasers in the other axial direction.

Figure 5.3   Laser sweep and object detection

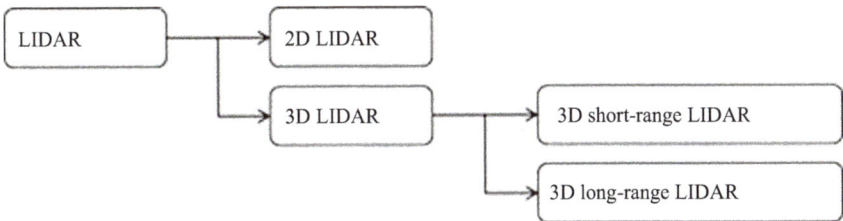

Figure 5.4   LIDAR system classification

The 2D LIDAR projects the laser from its installed position above the road toward the road surface and measures vehicles passing through the cross section. It is basically short-range measurement, and the longest distance is from the installation position to the road surface. Meanwhile, since the 3D projects light diagonally with respect to the road surface, the distance to be kept in the field of view (FoV) is more than the 2D type. Here long-range measurement is required. However, it is technically difficult to measure the long range because it is necessary to increase the laser intensity and increase the light-receiving sensitivity. There are only a few LIDAR products capable of performing long-distance measurement applicable to major intersections that have a large traffic volume. The short-range type LIDAR is mainly intended for an accurate measurement of a vehicle at short distance, for example, a vehicle profiling is assumed. Characteristics for each type of LIDAR system are summarized in Table 5.1.

Compared with on-board LIDAR system such as Velodyne HDL-64E, etc., there is a difference in the FoV design. The on-board LIDAR system is preferred to have 360 degrees horizontal FoV because it is needed to measure the surroundings in advance to the right or left turn. On the other hand, since the LIDAR system for intersection measurement has a fixed direction desired to see, it is preferred to narrow down the horizontal viewing angle to some extent and instead to measure the specific direction at higher point cloud density.

It is sometimes difficult to cover the entire intersection by single LIDAR system because the FoV is not wide enough, or there are obstacles in the intersection, and occlusion is unavoidable. Extending FoV may be required using multiple LIDAR systems. FoV is categorized as shown in Figure 5.5.

Table 5.1   Characteristics for each type of LIDAR system

	2D LIDAR	3D LIDAR Short-range type	3D LIDAR Long-range type
Measurement range	A few meters	A few tens of meters e.g. 1–30 m	A few hundreds of meters e.g. 1–200 m
Use case	Vehicle counting Shape profiling ([a])	Shape profiling Vehicle tracking (small area)	Vehicle tracking Vehicle queue measurement
Field of view		Small → Large	
Cost		Low → High	
Example	SICK product (no image)	Leddar Tech product (no image)	
	SICK	LEDDARM16	IHI 3D laser radar

[a]Since there is no information on the speed of the vehicle, the vehicle length is an estimated value.

Continuous FoV is a state in which it is continuous. It corresponds to an FoV measured with a single LIDAR (single monolithic FoV) or integrated FoV by combining overlapping views of each other from multiple LIDARs (continuous combined FoV).

Continuous combined FoV has technical issues such as how to accurately transfer vehicle ID and vehicle motion information between sensors within predetermined latency. Even if the same vehicle is being measured, since the measurement surface of the object and the measurement timing are different, the measurement results do not exactly match, so assumption-based integration is required in a software layer.

Discontinuous FoV is a state in which FoVs measured by a plurality of LIDARs are not continuous. Multiple 2D LIDAR system is often used in the case. Only the vicinity of the entrance to the intersection and the exit are measured.

As the discontinuous FoV includes sections that have not been measured, the issues presented in continuous combined FoV become more difficult. There is also an issue that it is impossible to detect stagnation such as accidents or parking in an intersection.

It is based on case by case, which is suitable to cover the area with a plurality of low-cost short-range LIDAR system in a FoV-combined manner or with single long-range LIDAR system. The short-range LIDAR system certainly comes at low cost, but it takes more installation and management costs to cover the same FoV. It is necessary to judge which is better by the overall system cost by taking into account the geographical conditions of the intersection.

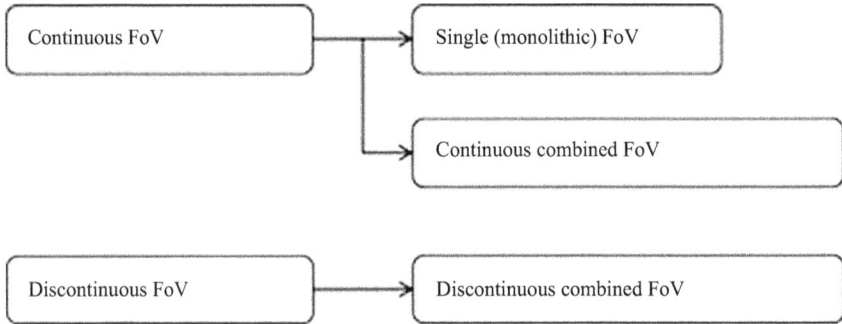

Figure 5.5 Type of field of view (FoV)

5.1.5 Performance of LIDAR system

Key performance indices are summarized in Table 5.2.

The specifications of major LIDAR systems are summarized in Table 5.3.

5.1.6 Current deployment status

Several cases have been reported about intersection monitoring using LIDAR system in the research phase and small-scale trial phase. However, in the social demonstration phase, it is quite few. One representative case is the one conducted with the Singaporean government using IHI's LIDAR system. This example is described in the following section.

5.2 Application of vehicle detection by an IHI's 3D laser radar

5.2.1 Practical application of a 3D laser radar is close at hand in playing a central role in the Intelligent Transport Systems

Intelligent Transport Systems (ITS) mediates the exchange of information among vehicles, roads, and people in order to prevent car accidents, avoid traffic jams, and address environmental problems and other challenges. After performing repeated demonstration experiments in Japan and Singapore, a practical application of the IHI's 3D laser radar for preventing accidents at intersections is close at hand.

5.2.2 Eyes that tell vehicles the road conditions at a nearby intersection

There are many kinds of ITS around us. Examples in Japan include the Vehicle Information and Communication System (VICS) as well as the Electronic Toll Collection System, and Driving Safety Support Systems (DSSS) that transmit information to vehicle-mounted navigation systems to inform drivers of road conditions like traffic jams and lane closures. In a DSSS, information is sent from infrared beacons to on-board units (OBUs, VICS-compatible car navigation units)

Table 5.2 Performance index

Index	Description
Maximum range	Maximum distance the LIDAR system can measure Beyond that, reflected light cannot be measured and the data will be lost The maximum distance is determined by measuring how many meters away it can detect the reflected light from an object with a specified reflectance
Field of view	Measureable area often specified by horizontal and vertical angle of view Example: 90 degrees (horizontal), ± 30 degrees (vertical)
Point cloud density	An index indicating how many points of measurement are in the unit area Example: 1 measurement point in every 0.3 degree in horizontal direction
Frame rate	An index indicating how many times field of view are updated per second
Beam profile	Spot shape of light projected toward objects If it is broad enough, there will be less detection leakage between each neighboring measurement point, but the amount of reflected light will be reduced which makes the maximum range shorter
Sweep method	A method indicating how to move the projection axis within field of view
Multi-hit time measurement	A method how to respond when multiple reflected lights return Light projected in one direction may come back several times due to dust in the air or the Like
Point cloud clustering method	A method of grouping points belonging to the same object
Vehicle detection capability	Percentage of successful vehicle detection For safety applications, it is necessary that the performance of object recognition is stable and guaranteed (functional safety)

Table 5.3 Specifications of major LIDAR system

	3D laser radar (IHI)	M8 (Quanergy)	Vista (Cepton)
Range	10–200 m	1–150 m *1 *1 : with 80% reflection object	1–200 m
Field of view	90 degrees (horizontal) ±/30 degrees (vertical)	360 degrees (horizontal) −17 to +3 degree (vertical)	60 degrees (horizontal) 24 degrees (vertical)
Distance resolution	(Non available)	3 cm	2.5 cm
Angular resolution	About 0.2 degree (horizontal)	0.03–0.2 degree (horizontal)	0.2 degree (horizontal)
Frame rate	3.3 Hz	5–30 Hz	15 Hz
Size	569 × 274 × 340 in mm 16 kg	102 × 102 × 86 in mm 0.9 kg	89 × 64 × 102 in mm 0.8 kg

that display relevant illustrations on the monitor and make sounds to inform drivers of any hazards. In Japan, the National Police Agency took the initiative from July 2011, in putting such a system into practical use to provide a margin of safety to prevent traffic accidents. Moving beyond this conventional DSSS, academic, business, and governmental circles are developing pilot programs for the next-generation DSSS to enable OBUs to determine which information is necessary, depending on the operating conditions of their vehicles. One such initiative is led by the UTMS (Universal Traffic Management Systems) Society of Japan. The 3D laser radar developed by the IHI (hereinafter called "laser radar") is employed at the core of equipment for facilitating safe driving mainly at intersections.

Meanwhile, a pilot program is being spearheaded by the Cabinet Office. This Cross-ministerial National Project for Science, Technology and Innovation is striving to develop an automated driving system (for autonomous cars) as a part of the Strategic Innovation Promotion Program. A laser radar also serves as the eyes for these autonomous cars, conveying information from intersections to them. The first step for achieving automated driving is the development of a system for sharing vehicle-to-vehicle (V2V) information among vehicles to prompt appropriate driving control. For instance, an automated breaking system that reacts when the distance between cars drops below a certain threshold so as to avoid rear-end collisions has already been put into practice. An additional infrastructure-to-vehicle (I2V) system is needed to communicate road conditions to vehicles to prompt appropriate driving control. Such a system judges how to direct vehicles that are approaching an intersection. The IHI's laser radar is at the core of the latter system that provides information on vehicles and pedestrians at intersections.

In fact, the IHI's laser radar has already been practically applied to a system that detects obstacles in railway crossings and warns approaching trains (refer to "3-D Laser Radar Level Crossing Obstacle Detection System," IHI Engineering Review Vol. 41, No. 2, pp. 51–57). For this purpose, more than 1,600 laser radars have been installed in Japan and 127 units installations are ongoing in Italy. In recent years, the IHI has been delivering 200–300 units annually, counting those sold both in Japan and abroad. The development of a laser radar for ITS was commenced at around the same time as that for railway crossings. Practical application is finally on the horizon after pilot programs both in Japan and abroad.

5.2.3   Instant identification of objects with reflected laser light

Simply put, a laser radar is a device for quickly scanning and monitoring a certain space. The laser light is irradiated onto the road surface while scanning in horizontal and vertical directions to calculate the distance to each irradiated spot by measuring the time it takes for the light to be reflected back to the unit. It takes about 0.3 s to cover the space starting from a stop line and moving across an intersection three or four lanes in width and about 150 m in depth. Constant monitoring of an intersection in this manner makes it possible to gauge the height and width of any object that enters the space and how fast it is approaching the intersection. A program written by IHI determines whether the approaching object is a car, a motorcycle, or a pedestrian (Figure 5.6). In addition, the movement is captured and traced from the moment the object enters the monitored space to send a signal to represent the movement in real time. If a vehicle with a DSSS compatible on a board device approaches an intersection with this system, the on-board device receives the information from the laser radar to display an alert according to the designed standards. The display will appear differently, depending on the manufacturer of the navigation system. But as far as the voice alert is concerned, a unified standard is applied by organizations and companies involved in ITS in order not to confuse drivers with different kinds of voice guidance.

5.2.4 Advantage of all-weather capability and fast data processing

Reportedly, most intersection accidents involve accidental contact between vehicles turning right and those advancing straight, or pedestrians crossing intersections getting hit by vehicles turning right. During a right turn, the driver of a bus or large truck sometimes experiences difficulty noticing the vehicles in the oncoming lane. Even in such a situation, the system monitors any movements in the oncoming lane to facilitate the right turn. The system informs the driver of any movements of bicycles or pedestrians on the crosswalk the vehicle is approaching after turning right.

A position for deploying the laser radar at the intersection and in the direction for monitoring the road depends on the kind of information service intended by a service provider. Radar positions and directions are determined after traffic analysis, accident analysis by the police, and on-site surveys, but they are normally installed to provide vehicles traveling on busy roads with traffic information.

Video images from a video camera offer another way to monitor space. Unfortunately, these cameras are often incapable of measuring distances or capturing accurate images during certain hours of the day or types of weather, including nighttime and rain. In this regard, a laser radar is more adaptable to surrounding environments. In short, the advantages of a laser radar comprise a

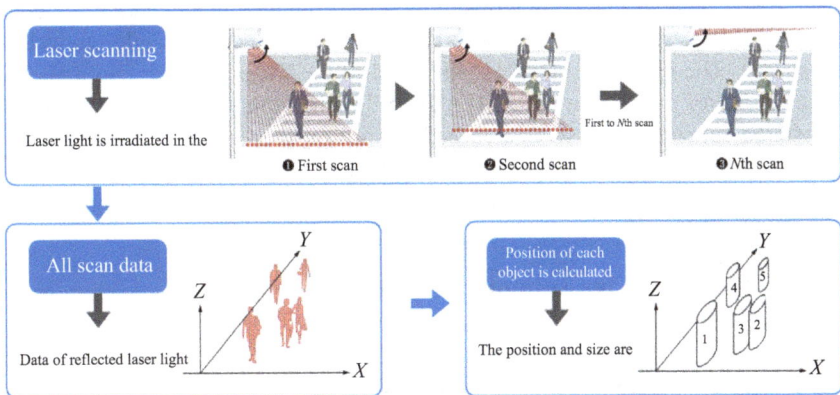

Figure 5.6 Method for detecting vehicles and pedestrians

wider scanning range, faster scanning, and smaller files sizes than video for recording and distribution, which enables a simple device with the processing power of a PC to make appropriate real-time judgments without depending on a huge server. Incidentally, the evaluation software was also developed by the IHI. This original software selectively processes the data that a service provider needs instead of processing all information.

5.2.5   Pilot program in Singapore

The IHI has introduced a technology attaché system, in which junior and mid-career engineers are stationed overseas for an extended stay to engage in marketing research, explore useful technologies, and conduct joint research with local public institutions and universities so as to plan a project leading to new business in the future.

Prior to the establishment of the IHI Asia Pacific in April 2012, a technology attaché has been deployed to Singapore since October 2010 in addition to New York and London.

Singapore has been dubbed "the world's testing ground." The advanced information society with a tiny territory that fits inside the Yamanote Railway loop line in Tokyo attracts many pilot programs by companies from every corner of the world. The government is known to offer generous support to such experiments. Thanks to the local attaché, the IHI signed a comprehensive research and development (R&D) agreement with the A*STAR (Agency for Science, Technology and Research) of Singapore. The partnership extends across the three areas of information and communication, production technologies, and environmental science and engineering. A demonstration of ITS technologies was commenced in December 2012 as a part of the partnership. In Phase 1, the on-site survey and system design for a six-lane intersection in Jurong district was followed by the installation of two laser radars. Data was then accumulated to check if these radars reliably detected target objects (e.g., vehicles, motorcycles, and pedestrians), to make sure that there was no mis-detection and that detection is not affected during certain hours or types of weather, such as nighttime, rain, and so forth. Taking advantage of the left-hand traffic like in Japan, radars were examined in terms of their effectiveness in preventing accidents involving vehicles turning right and other vehicles advancing straight, as well as pedestrians. The pilot program was continued in the fiscal year 2015 to make the service a reality in the near future (Figure 5.7). The company intends to enhance the offered functions and services based mainly on the accumulated data and to advocate the wider application of DSSS in Japan and beyond.

The specific contents of the two actual projects conducted in Singapore are shown in the following sections.

5.2.5.1   Safety driving support for junction

Introduction

In urban cities, a high percentage of road accidents happen at traffic junctions where vehicles with conflicting movements get involved in collisions [1]. The analysis shows that most of these accidents take place due to inadequate surveillance and/or erroneous decision-making by the drivers involved. Of those collisions

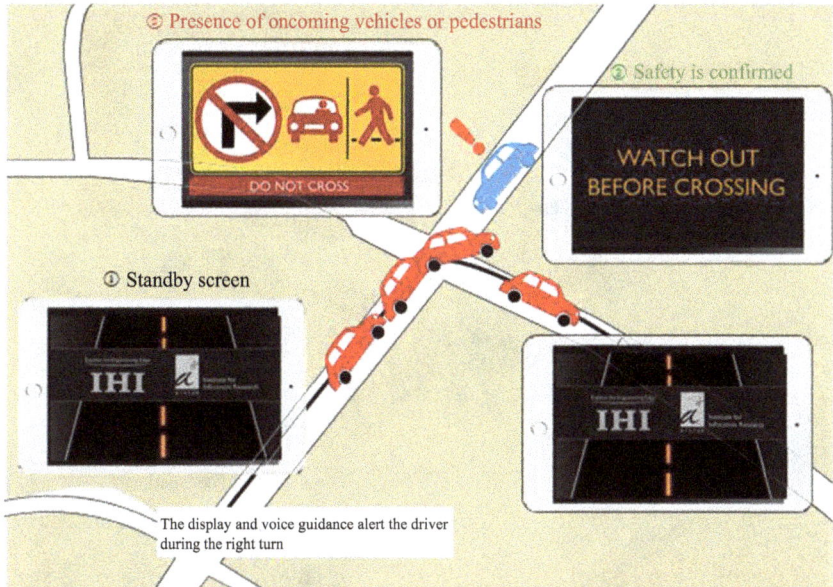

Figure 5.7   Example of display for facilitating safe driving (pilot program in Singapore)

at traffic junctions, it has been observed that the two most common scenarios are the right-turning and crossover scenarios. In countries with left-hand traffic, a right-turning collision involves a vehicle having minor phase priority for turning right and another vehicle having major phase priority for going from the opposite direction. The crossover scenario involves a through-moving vehicle that runs a red light and collides with another vehicle from a perpendicular approach.

Our Sensors-Augmented INTersection (SAINT) project aims to develop a system that mitigates these accident scenarios. The target system leverages on advanced sensory and communication technologies to anticipate and warn drivers of dangerous situations. The technologies being used in the SAINT project include vehicle and pedestrian localization with a 3D laser radar and GPS, and vehicle-to-infrastructure message exchanges based on dedicated short-range communications (DSRC).

This chapter follows our previous presentation on the design of the SAINT system and preliminary results of some of the subsystems [2], by covering the actual implementation of the whole system on road conditions in Singapore, together with the evaluation of the reliability and accuracy of the system.

System overview

Figure 5.8 shows the overview of the system. The system consists of the 3D laser radar, OBU in the vehicle, and the main controller.

The 3D laser radar (developed by IHI Corporation) refers to the LIDAR sensor. The sensor measures the existence, position, size, and speed of vehicles and pedestrians located within the target junction.

The OBU has a display to show the warning massage to a driver. The OBU is equipped with a GNSS module to track the position of the targeted vehicle.

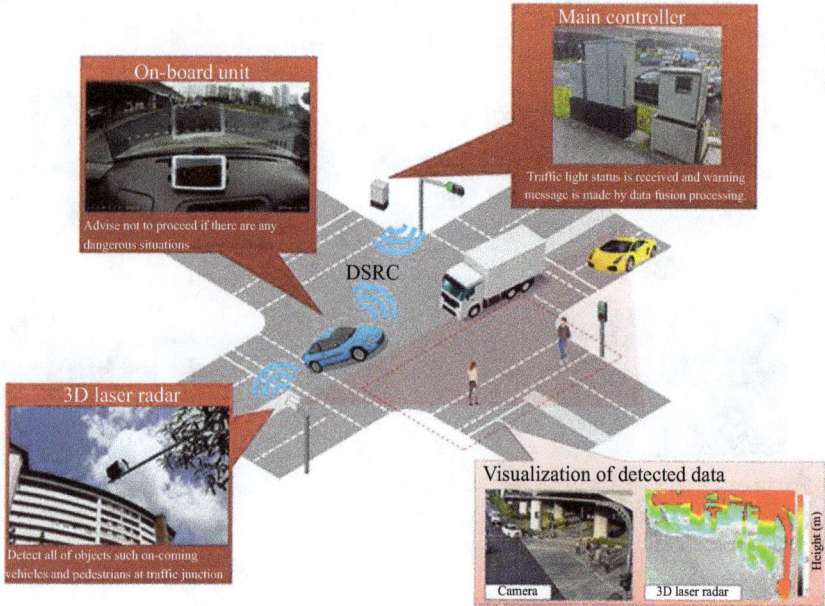

Figure 5.8   Overview of SAINT system

The main controller interfaces with the traffic light controller to monitor the traffic light status. The base station of the GNSS is also installed at the road side to improve the vehicular position accuracy measured by the OBU's GNSS module, using RTK-GPS technology. All of the sensory data will be processed with data fusion analysis within the main controller to reach a decision, and the warning message is sent to OBU.

The system provides the following driver assistance services:

1. avoidance of right turn collision;
2. avoidance of accidents with pedestrian crossing (left turn and right turn).

Field tests results

The system was installed at Jurong East Rd–Toh Guan Rd junction in Singapore, with field tests performed in different test conditions (off-peak, on-peak, rain, and night). In this performance test, each subsystem (3D laser radar, GNSS, and DSRC) and overall safety application performance were evaluated.

The test for vehicle-detection using a 3D laser radar indicates 100% detection accuracy for all leading vehicles extracted from the recorded videos. The detected location of the test vehicle matches well (within 50 cm on average) to odometer-recorded location of the test vehicle. Regarding the performance of the warning message, its accuracy for the right-turn scenario was 97.72%.

5.2.5.2   Traffic violation detection system

Introduction

Another important thing about traffic safety in urban areas is to reduce traffic violations. For example, illegal right/left turns, U turns, etc. are potential violations that can cause accidents, which often caused by these are actually seen in large junctions. Therefore, the reminding drivers of these violations and deterring violations can reduce accidents and reduce traffic congestion caused by accidents. To reduce traffic violations in Singapore, the automatic violation detection project led by Traffic Police has been in progress since May 2018.

This project is a system using a 3D laser radar and several cameras.

The system tracks the movement of the vehicle using a 3D laser radar and automatically detects the following three types of violation as shown in Figure 5.9: an illegal U turn, a vehicle turning left from a left turn prohibited lane, a stopped vehicle in a yellow box that it interferes to drive in the junction. A 3D laser radar is

Capturing three traffic offences

The new cameras use 3D laser technology to target the following offences.

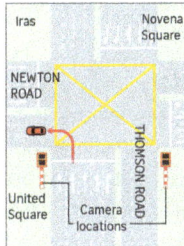

Turning in non-turning lanes

Offence
Failing to form up correctly when turning left under Rule 7 of the Road Traffic Rules

Penalty
• Composition fine of $130
• Four demerit points

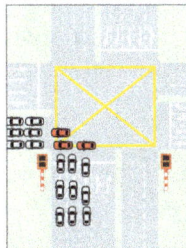

Remaining stationary in yellow boxes

Offence
Causing unnecessary obstruction to vehicles proceeding to or along the box junction by remaining inside it under Rule 26 of the Road Traffic Rules

Penalty
Composition fine of $70

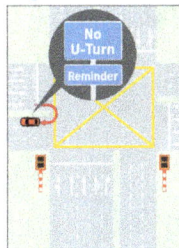

Illegal U-turns

Offence
Making an unauthorized U-Turn under Rule 13(1) of the Road Traffic Rules

Penalty
Composition fine of $70

STRAITS TIMES GRAPHICS

Figure 5.9   Violations to be detected [3]

adopted to raise the clearance rate because dangerous traffic violations occur often from evening to nighttime, and in general a 3D laser radar has a higher detection rate than a camera.

System overview

Figure 5.10 outlines the system. This system consists of a 3D laser radar, an auto number-plate recognition (ANPR) camera, and main controller.

A 3D laser radar measures all vehicles, positions, and speeds within the target junction as well as the safe driving support mentioned earlier.

The ANPR camera is connected to the main controller by Wi-Fi communication using wireless access point, transmits camera data, and recognizes the license plate of the vehicle as necessary.

Similarly, the main controller tracks the route of movement of the vehicle based on the 3D laser radar information and recognizes various kinds of vehicles in violation. After recognizing the offending vehicle, the controller will extract the license plate information of the offending vehicle from the ANPR camera and integrate the time information and will process it to identify the violation detection vehicle and be informed in detail.

Tracking technology of a 3D laser radar

In applying this method to detect violation, the technical problems aspect of the 3D laser radar is that it is necessary to keep track of the same vehicle. In "safe driving support for junction" mentioned earlier, the function was realized by judging whether or not there is a target vehicle, but in the case of detection of violation, the 3D laser radar must continue to track the same vehicle until it recognizes as a violation. If this fails, there is also the possibility that the violation detection rate decreases or the normal vehicle is mistakenly recognized as a violating vehicle as the worst case.

In addition, the intersection which is the object of violation detection is relatively wide. It is necessary to capture vehicles far away as on the main road there

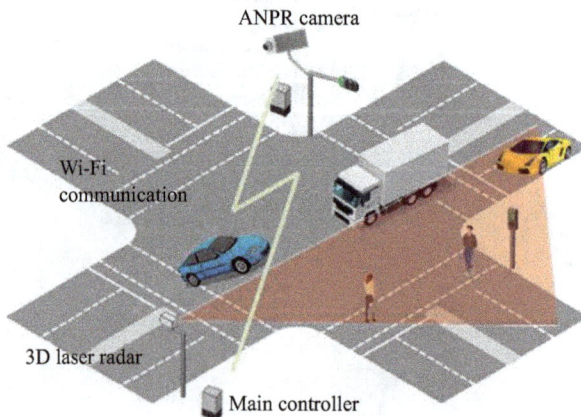

Figure 5.10   Overview of system

are many large vehicles such as buses, and the vehicles to be tracked may get temporarily hidden, which is one of the challenges.

In our 3D laser radar, in order to enable tracking even at such a wide intersection, we improved the tracking rate by implementing separation/combination judgment processing of highly accurate vehicles, processing to preserve tracking by presuming that the vehicle is there even when occlusion (optical "hiding" by objects) occurs.

Field test

This time it will be tested for 3 months at the intersection of Thomson Road and Newton Road called Novena Junction. The installation of the system has already been completed, which is currently under test, and we acquire actual violation detection data and brush up processing to achieve the target violation detection rate.

Venturing into the world with ITS that embodies smart technologies of Japan

The ITS pursued by Japan mediates the exchange of information among vehicles, roads, and people to build a safe traffic system. As mentioned earlier, V2V systems, including those preventing rear-end collisions, are almost a commercial reality thanks to technologies developed by automobile companies. Communication between I2V with a laser radar as featured here will also be put into practical use in the near future, mainly at intersections. Unfortunately, no system has been developed for alerting pedestrians about approaching vehicles although a detection system with a laser radar at an intersection can already inform vehicles regarding the presence of pedestrians. Some automobile manufacturers are developing applications using smartphones, to alert pedestrians. But they need to accumulate and examine data related to traffic safety. Accordingly, services are being explored to reliably inform pedestrians of approaching vehicles and to provide information to regulate traffic signals to prevent the elderly and other vulnerable people in traffic from being stranded in intersections.

R&D of ITS is conducted through the partnership of industry, academia, and the government. The government has also set a national strategy to promote infrastructure and related services as a package, not only in Japan, but also throughout the world. The package would include technologies for automated driving, traffic sensing, and the safety of pedestrians. The laser radar made by the IHI is expected to continue to play a core role as a reliable device to make that possible.

Note:

Terms such as "turning right" included in this report are based on countries where traffic drives on the left-hand side of the road (e.g., Japan, Singapore, etc.).

The systems in this report can also be applied in countries where traffic drives on the right-hand side of the road.

References

[1]   Eun-Ha C. (2010). Crash Factors in Intersection-Related Crashes: An On-Scene Perspective, Report DOT HS 811 366.

[2]   Ang C. W., Zhu J., and Hoang A. T. (2014). Enhancing Junction Safety With Connected Vehicles, ITS Asia Pacific Forum 2014.

[3]   The Straits Times (2018). Traffic Police to trial new 3D laser cameras at Thomson-Newton junction black spot https://www.straitstimes.com/singa-pore/traffic-police-trials-new-cameras-at-thomson-newton-junction-black-spot.

Chapter 6

Vehicle detection at intersection by RADAR system

Yoichi Nakagawa[1]

6.1 Background

Millimetre-wave radar is mainly used as a vehicle onboard sensor for monitoring the front of a car and contributes to the spread of ADAS (Advanced Driving Assist System) such as Active Cruise Control and Pre-Crash Safety. In addition, it is considered that when compared with visible camera and LiDAR (laser imaging detection and ranging) that have the similar sensing function, it is superior in environmental robustness and velocity measurement. Furthermore, since it can be mounted inside the emblem or bumper, the flexibility of installation and design is also a feature of millimetre-wave radar.

The millimetre-wave indicates radio wave in frequency range of about 30–300 GHz, and it is called as such since it is the frequency band in which the wavelength becomes 1 cm or less. Radar is an active sensor and leverages the physical phenomenon in which the radio wave to be transmitted is scattered in many directions on the surface of the object. It receives the echoes of the scattered wave and measures the distance to the object, the velocity in range, and the angle of arrival.

As a physical feature, the millimetre-wave radar has a shorter wavelength than the radar using microwave bands (3–30 GHz), so the resolution of the Doppler frequency obtained with the same observation time increases. Further, it is possible to reduce the size of the antenna that achieves the required angular resolution in proportion to the wavelength. Actually, the feature of the millimetre-wave, which makes the radar device compact and thin, is an important appeal point as a sensor part for automotive applications.

In terms of standardization of frequency use, as response to the allocation of 76–81 GHz band to the radio location service (radar) as the international radio law prescribed by the ITU-R [1], in the United States and other countries, legislation activities on using this frequency band for automotive radar have been proceeding. The radar using the several GHz band can achieve high separation performance theoretically with the range resolution in the order of centimetre. Such improvement

[1]Panasonic Corporation, Osaka, Japan

in resolution leads to expansion of the detection function of millimetre-wave radar in public road space where not only vehicles but also pedestrians and motorcycles come and go.

The 79 GHz radar indicates a high-resolution short-range radar using 77–81 GHz, and it can be used by acquiring technical certification as specific low power radio in Japan. At the present time, industrial standards of Japan and Europe are defined in the 79 GHz radar as distinguishing the 77 GHz radar using 76–77 GHz which is already spread as automotive radar [2,3].

From the viewpoint of practical application, securing device performance by a semiconductor manufacturing process suitable for mass production leads to cost reduction, so that circuit design technology of CMOS (complementary metal–oxide–semiconductor) has been developed even in the millimetre-wave band [4]. Now, the radar using the 76–81 GHz band is realized as one chip device by CMOS as a mixed system of analogue circuit and digital circuit. As a background to such aggressive investment for millimetre-wave devices, there is a global competition for developing automated driving systems, and the high-resolution radar becomes a key sensor technology for recognizing the road space.

6.2   High-resolution millimetre-wave radar

Expectations for spatial imaging by millimetre waves are increasing, and the way of scanning a short range with a wide angle is becoming a standard operation of the high-resolution 79 GHz band radar. The effective performance of this radar is about 30–40 m for pedestrians and about 70–100 m for auto-mobiles as the maximum detection distance depending on the target RCS (radar cross section). In the case of a typical planar antenna configuration, the FOV (field of view) at which the location of target is detectable is about $\pm 45°$, but the FOV at which just the range is measurable is up to $\pm 75°$ (see Figure 6.1). Further, since the usable frequency bandwidth exceeds 1 GHz, the theoretical range resolution becomes 15 cm or less. Therefore, scattering points existing in the same range bin are reduced, and the angular resolution is effectively improved.

Also, since a Doppler frequency shift occurs in the echo of a moving target as the physical characteristic, the velocity of the target can be estimated from the Doppler frequency. When the carrier wave is 79 GHz, the wavelength is about 4 mm; for example, a velocity resolution of 1 m/s is obtained with an observation time of 2 ms. Such high-range resolution makes the radar excellent in separation performance for multiple targets with slight velocity difference and close to each other. In the case of the pedestrian detection, it is possible to analyse the Doppler frequency spread due to the vibration of limbs as a feature value superimposed on the Doppler frequency shift. Indeed, understanding echo char-acteristics inherent in such targets leads to the extension of the radar detection function [5].

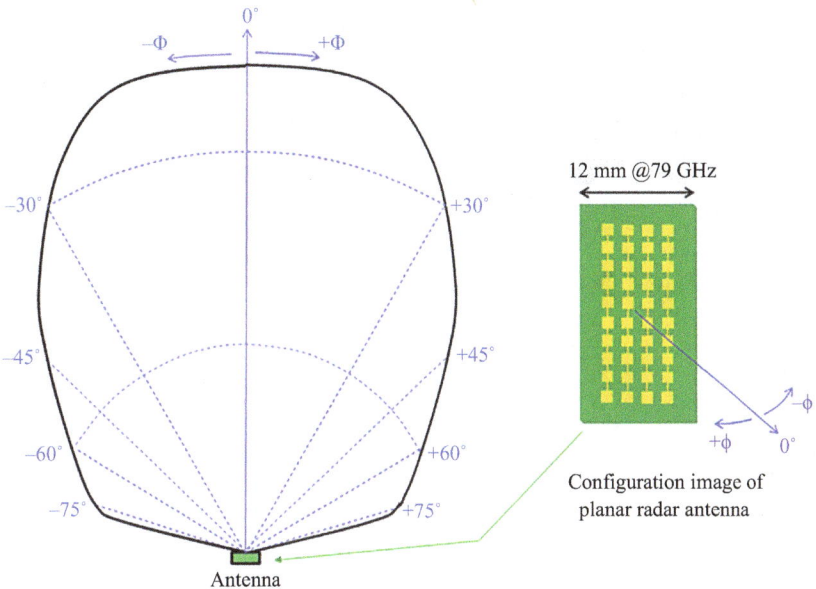

Figure 6.1 Detection angle range based on directivity of a typical planar antenna

Moreover, as the device technology for compensating variations in the RF circuit by the digital signal processing evolves, it is possible to measure the phase of the received signal even with millimetre waves. In particular, the angle of arrival estimation based on the phase difference between the array antenna elements can be realized with higher accuracy. In addition, the MIMO (multiple-input and multiple-output) radar, which is virtually composed of the array antenna, can increase the angular resolution while suppressing the number of antenna elements.

Here, the output data of the radar device is defined as a scan data of mesh-like cells, the physical domains of which are distance, angle, and Doppler velocity. This is multidimensional information reflecting the resolution of each region and is a point cloud data, including not only echoes of moving objects but also stationary objects and possibly echoes scattered by raindrops at the same time (see Figure 6.2). As the radar device assumed to be used for onboard sensor applications, the specification of the scan period, that is the data update period, is typically 50–100 ms.

In order to improve the effective detection accuracy of the radar system, it is essential to optimize clustering and tracking based on the echo characteristics of targets (see Figure 6.3). For example, the echoes of a vehicle observed with the high-resolution radar are separated into a large number of scattering points. Therefore, it is necessary to treat the spatially spread candidate cells as the same group by using the Doppler velocity and the like.

Echo power (dB)   Doppler velocity (km/h)

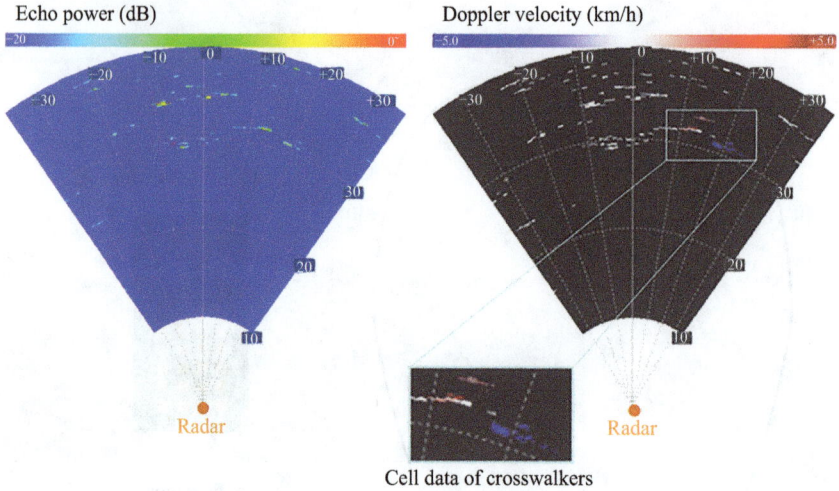

Cell data of crosswalkers

Figure 6.2   Radar scan data with respect to echo power and Doppler velocity

Before clustering process   After object detection

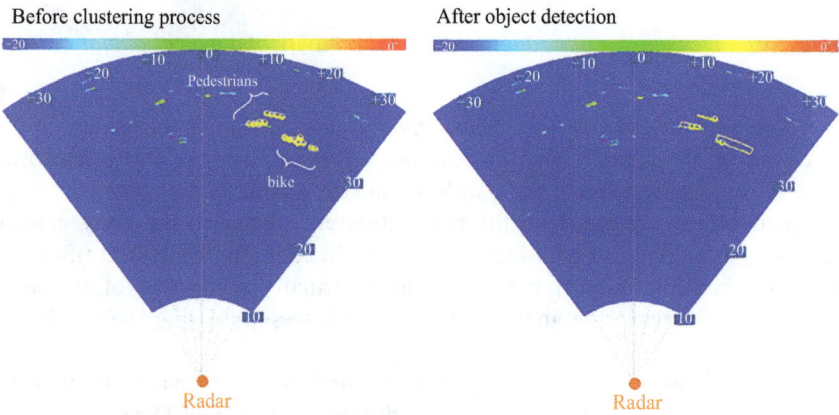

Figure 6.3   Radar data processing aiming to the target detection

6.3   Roadside radar system

The millimetre-wave radar having the performance and the function as described earlier is installed on the roadside of intersections and used for separating and detecting pedestrians and vehicles passing [6]. The purpose of setting the radar at the intersections is to determine the pedestrian presence and to measure the traffic flow, and it is considered to be applicable not only to safety support but also to traffic signal control and the like. When realizing the function of object recognition, visible cameras are generally used. However, with respect to detecting the presence of a pedestrian and counting the number of vehicles, the radar can obtain stable accuracy regardless of weather or time.

For such a roadside sensor, it is required for installation to avoid the occlusion caused by the vehicle in front of the detection target. The height of traffic signal installed so as to ensure visibility from vehicle is one of the measures of the installation of the sensor. For example, when the installation height is 5 m above the ground and the vertical plane beam width of the millimetre-wave radar is 10°, the effective range coverage for passenger car is calculated to be about 15–75 m (see Figure 6.4). It indicates the sensitivity characteristics within the data frame at a period of 100 ms or less, when the radio link design is done under consideration of the regulation of 79 GHz band high-resolution radar [2].

In order to verify coverage in the distance direction, radar fundamental experiments targeting passenger cars, motorcycles, and pedestrians are carried out using a test course. This test course is an environment simulating a general inter-section, and a radar sensor unit was attached to a signal pole, and the sensitivity characteristics of each target with respect to the road surface distance were con-firmed based on the echo power. Actually, the echo power measured by the radar fluctuates depending on the direction and shape of the vehicle, but the difference between RCS for passenger cars and motorcycles can be regarded as about −6 dB.

Regarding the passenger car, the positioning performance of the millimetre-wave radar installed on the roadside is confirmed, under the condition that multiple cars run on two lanes. Specifically, coverage and accuracy are verified by esti-mating the lane travelling from the instantaneous measurement data with the radar installation point as the distance reference. The driving scenario of this verification experiment is set so that two passenger cars travelling parallel to each other on

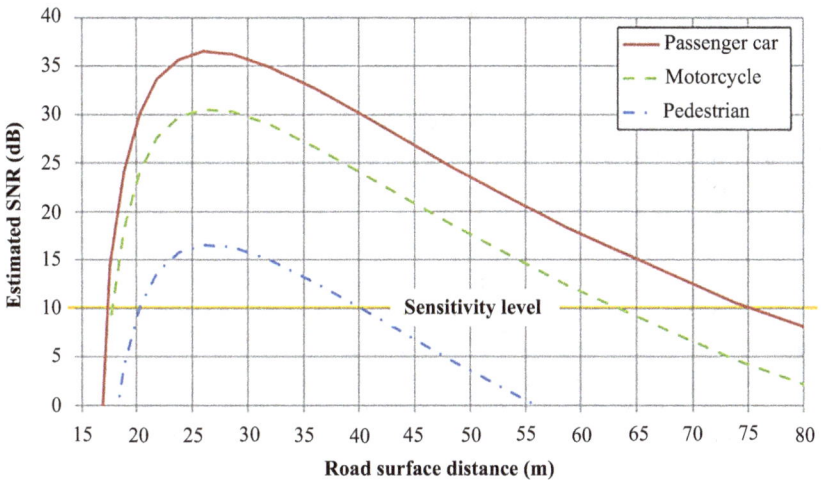

Figure 6.4   Sensitivity property of the radar targets on surface distance

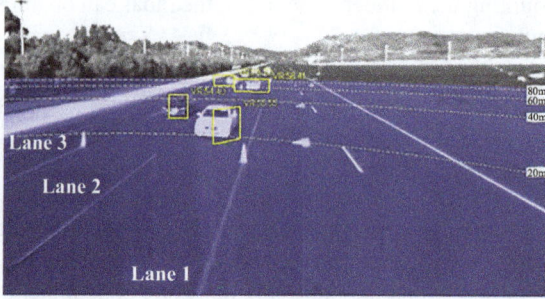

Figure 6.5   Snapshot of the radar data for parallel driving cars as superposed to picture

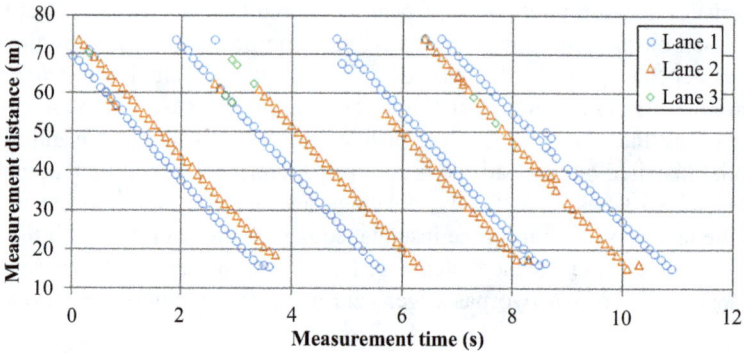

Figure 6.6   Distance estimation result at each lane for multiple sets of parallel running cars

adjacent lanes successively come from outside the coverage and pass by the side of the installation point. Each vehicle keeps the vehicle speed equal to 60 km/h and keeps the distance between the front and rear cars for 2 s, that is about 30 m converted into the distance (see Figure 6.5).

In Figure 6.6, the lane acquired for each measurement frame of the radar and the estimated value of the distance are classified and plotted for each lane. Focusing on the plot of the specific lane, the difference in the direction of each axis of the graph indicates the distance between vehicles, and when the distance is 50 m or less, positioning is possible with high accuracy. On the other hand, when the distance exceeds 50 m, undetected due to occlusion of the preceding vehicle and overdetected by multipath occurrence due to the parallel running vehicle are observed. However, the coverage of passenger cars with respect to two lanes has been secured in the range of about 15–75 m, and the performance according to the link design mentioned earlier is obtained.

6.4 Technical verification under severe weather condition

6.4.1 Objective

For radars where stable operation is expected even in bad weather, heavy rainfall is the most severe condition. Even in the millimetre-wave radar that scans the short range, under the heavy rainfall environment, it is unable to ignore the influence of radiating radio waves attenuating due to scattering of raindrops. In contrast, even if the amount of snowfall is very large, the actual precipitation does not become so much. Further, radio waves are absorbed by water, but snow is ice grains, the attenuation level of millimetre-wave propagation is relatively low at snowfall. In other words, in order to put the radar into practical use, technical verification assuming hard rain caused by typhoons and the like is required.

In radio equipment using high frequencies such as millimetre-wave band, when a water film covers the housing to heavy rain, the attenuation occurring there is large, which is a cause of sensitivity deterioration. Also, when snow falls on the housing due to blizzard, it will be covered with a thick water film at the time of snow melting, which is a factor of deterioration in the same way. Therefore, in order to ensure weather resistance as a roadside radar, a measure to prevent the water film from occurring on the surface of the radome, such as the equipment of an eave, is required. Furthermore, in snowy cold climates, it is also a measure to install a heater so as to avoid becoming easier to cover with snow on the radome cooled by cold wind.

6.4.2 Design for heavy rainfall condition

In the following, it explains the radio link design of the radar system based on the radio wave propagation characteristics of the 79 GHz band millimetre wave in the rainy environment. In the link design of the radar system, the maximum detection distance based on the RCS of the object is estimated after formulating the attenuation amount with respect to the rainfall intensity. The effectiveness of the link design considering these rain attenuations has been confirmed by verifying the echo characteristics and detection performance in weather conditions, including actual storm and blizzard.

As the initial study, propagation experiments using a 79 GHz band radar were carried out in a large test facility capable of keeping the rainfall intensity constant in order to quantitatively grasp the distance attenuation characteristics with respect to the rainfall intensity. Specifically, the echo power was measured while varying the rainfall intensity under the condition of a separation distance of 40 m by using a standard reflector as the target of the radar. By analysing the propagation attenuation caused by rainfall, it is confirmed that as the rain intensity increases, the average value of attenuation and its dispersion tends to increase together (see Figure 6.7). Here, by installing the eave so as not to generate water film on the surface of the radome, the influence of water drop adhesion is kept to a negligible level.

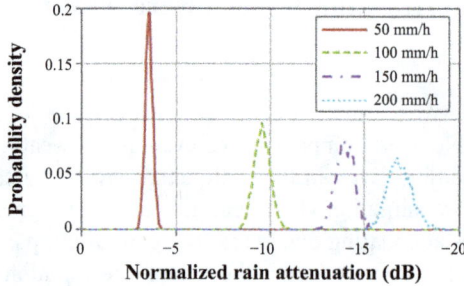

Figure 6.7   Measurement data on rain attenuation at the distance of 40 m

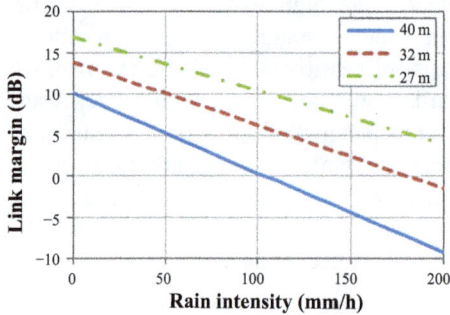

Figure 6.8   Estimation value of the link margin with respect to rain intensity

By modelling the propagation attenuation characteristic with respect to the amount of rainfall based on the result of the experimental analysis, it can be reflected in the radio link design of the 79 GHz radar system. Specifically, based on the experimental data of rainfall environments, the relational expression concerning the rain intensity and the attenuation per unit distance is obtained. For example, when the line margin for detecting a pedestrian ahead 40 m is designed to +10 dB, it is possible to estimate the margin is +5 dB at the rain intensity of 50 mm/h and 0 dB at 100 mm/h. Alternatively, it is also possible to estimate the effective coverage such as a distance of 32 m at which the link margin of +10 dB can be secured at the rainfall intensity of 50 mm/h and 27 m at 100 mm/h in the same way (see Figure 6.8).

6.4.3   Experiment in snowfall field

In cold climates where the amount of snowfall is large, snowstorms sometimes cause poor visibility, so the use of radar is particularly expected compared to visible camera and LiDAR. As described earlier, if radar technology capable in severe rainfall conditions is established, robustness in snowfall environments can be relatively easily secured.

In the following, field experiment results of the radar using 79 GHz band conducted in a heavy snowfall region are explained.

In the field experiment, a pillar was temporarily installed on the roadside of a single road, and a set of radar prototype equipment was installed. In order to detect a vehicle travelling behind a large-sized vehicle without missing, the installation angle is adjusted so as to be irradiated by the radar device from the backwards of the vehicles (see Figure 6.9). Moreover, the measurement accuracy was verified as vehicle counter by implementing the function of counting the number of various vehicles passing through to the radar detection software.

Several consecutive 24-h data, including severe weather conditions such as snowstorm, are selected for verification of radar measurement accuracy. Table 6.1 shows the accuracy verification result of the vehicle counting for each data of a typical snow and a heavy snow day and a day, including sleet and snowstorm. Although some miscounts occurred, in these cases, the detection accuracy of 99 per cent or more was obtained. Besides, the same functional verification was carried out by extracting all the time periods in which the vehicle does not present for more than 2 min, but no false alarm had occurred in either data.

When analysed in more detail, there are cases when the number of large vehicles is over-detected or when two vehicles shortening the inter-vehicle distance are counted as one. The occurrence of these events is not attributable to weather conditions such as rainfall or snowfall. Since the suppressing over-detection of a large vehicle becomes a trade-off with the separation performance of vehicles in proximity. In other words, if parameters are adjusted so as to suppress excessive detection of a truck or the like, less detection in multiple vehicles approaching close to each other tends to increase. In practice, algorithms and parameters will be optimized to meet the performance requirements required by roadside sensor applications.

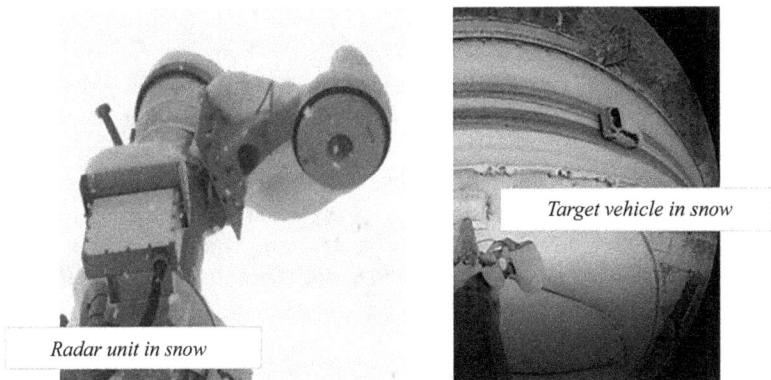

Figure 6.9   Roadside radar experiment in the snowfall field

Table 6.1   Verification results of vehicle count accuracy in snowfall environments

Weather condition	Number of vehicles (unit)	Overcount (unit)	Uncount (unit)	False alarm
Typical snow	3,221	2	0	None
Heavy snow[a]	4,598	1	1	None
Wet snow and snowstorm[b]	2,930	4	0	None

[a]Amount of snow 210 mm/day.
[b]Amount of snow 80 mm/day and maximum wind speed 14 m/s.

6.5   Detection accuracy verification on public road

We have addressed on the research initiative to apply the latest radar technology to infrastructure systems in consideration of evolution of millimetre-wave device and sensor software. Specifically, assuming ITS (Intelligent Transport System) application such as cooperative driving support, we have been promoting technology development and demonstration experiment aiming to apply high-resolution millimetre-wave radar to roadside sensor at intersection [6,7]. In order to provide so-called look-ahead information as the driving support, an appropriate wireless communication technology that shares sensor data in real time will be used in addition to sensor technology that accurately measures the speed of approaching vehicles and the position of crossing pedestrians.

For this application, it is necessary to acquire the position and speed of the detection target with high precision in real time, and it is expected that the high-resolution radar technology will be deployed while suppressing cost. From another viewpoint, for example in traffic signal control applications, it is required to measure direction-specific traffic flow at the intersection. In Japan, the traffic flow of public roads is measured with ultrasonic sensors installed overhead of each lane. In other words, if the millimetre-wave radar installed on the roadside for the driving support also realizes the function of measuring the number of vehicles passing through multiple lanes, the total cost as infrastructure equipment can be suppressed.

Currently, the experimental system of 79 GHz band radar is installed at an actual intersection of public road, and the detection of pedestrian on the crosswalks and the measurement of traffic volume on the inflow paths are verified. That verification field is a point where a road of one lane on one side intersects a main road of two lanes which is often seen in public roads in Japan, and the environment is high in traffic volume despite the narrow road width. Also, because it is located near a subway station, pedestrians and bicycles are increasingly passing in morning and evening time zones where the roads are also crowded.

The experimental system is installed in an existing signal pillar and consists of two radar units attached to about 5 m above the ground and a PC as control unit placed inside a cabinet. In this verification experiment, each of the radar devices is adjusted to be the diagonal direction and the inflow direction. Figure 6.10 shows

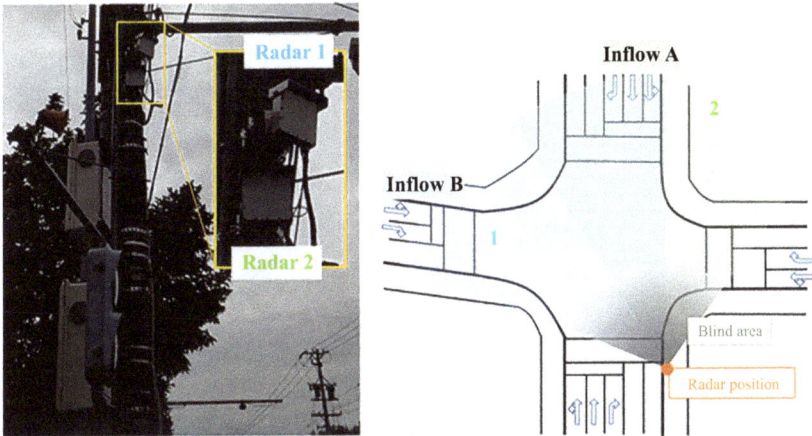

Figure 6.10 System configuration for the public road experiment

the positional relation between the detection range for pedestrian and each lane, and both radars are operated under the condition that the FOV on the horizontal plane is set to 70°. Since the vertical plane beam width of the radar antenna is about 10°, the vicinity of the ground level distance of about 15 m from the radar installation point becomes the blind area.

Here, the effective detection range of vehicles shows characteristics depending on the size of the vehicle body and the shape of its surface, except for the influence of occlusion caused by a large car or the like. Also, the measured positional accuracy of the vehicle also depends on the travelling direction and the relative distance of the target vehicle with respect to the radar. That is, since the echoes from the front and rear surfaces are dominant in the vehicle moving in the range direction, angle spreading is small and location identification is relatively easy. In the case of lateral movement or turning, unevenly spread echoes are observed from the side of the vehicle, so that ingenuity is required for clustering processing. Since this tendency is more prominent as the distance from the radar is shorter, the position accuracy does not necessarily degrade in proportion to the distance.

When measuring the traffic flow at the intersection, it is required to count the number of vehicles in three travel directions in which the vehicle goes straight or turns left or right. Efficient measurement is considered to capture the transit at the timing just after branching where the travel direction is determined; therefore, approach to count the number of vehicles in the intersection is taken. Specifically, after installing at the actual intersection, the travelling locus of the inflowing vehicle is grasped from the acquired data and the virtual lines of transit determination are set. Then, when the trajectory obtained by the radar tracking process crosses these determination lines, the number of each passing vehicle is counted up.

Regarding such counting function of the number of vehicles, verification experiments of measurement accuracy are carried out for two inflow paths with

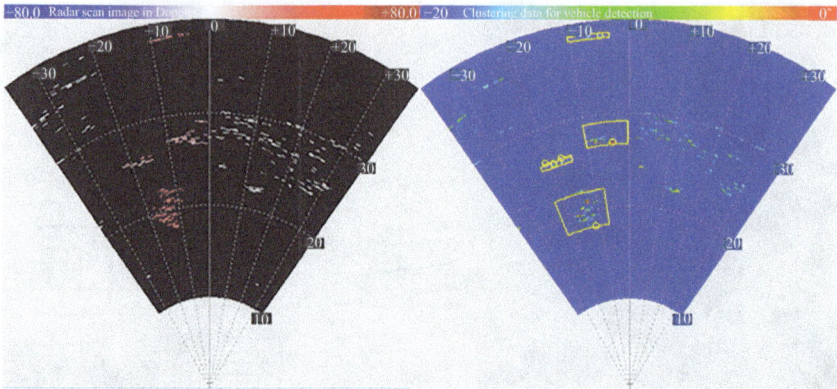

Figure 6.11   Sample of the radar measurement data in the verification of vehicle counter

Figure 6.12   Camera image taken at the same time of the above radar measurement

different travelling conditions. The target conditions in these verification experiments are the case of flowing into the intersection at three lanes from the front direction of the radar and the case of flowing in two lanes from the oblique direction (see Figure 6.10). For each inflow route, the number of vehicles passing through is acquired by radar system installed on the roadside for each of three directions of straight ahead, right turn, and left turn. As for the true value of the number of passing vehicles, it is acquired by visually recognizing the camera image recorded at the same time as the radar measurement (see Figures 6.11 and 6.12).

As a specific scene to verify measurement accuracy, a time zone of heavy traffic and heavy rainy weather is selected. In other words, in order to obtain verification results with high practicality, relatively harsh environmental conditions are set as performance evaluation of millimetre-wave radar. In order to statistically analyse errors in the radar measurement, the average value for about 60 min is calculated, using the signal cycle of each inflow path as one data sample (see Figure 6.13). Table 6.2 shows the short time measurement sample for the inflow A

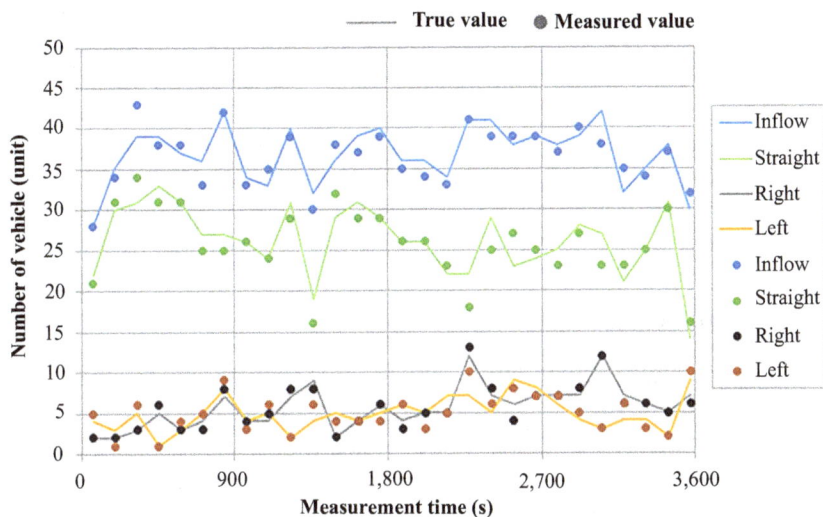

Figure 6.13   Radar measurement data of direction-specific traffic volume at three lane inflow

Table 6.2   Short time measurement sample of direction-specific traffic volume

	Run straight	Turn right	Turn left
True value (unit)	733	159	137
Measured value (unit)	724	159	141
Average of RMSE (unit)	2.1	0.8	1.3

Table 6.3   Evaluation results on the measurement of direction-specific traffic volume

		Run straight	Turn right	Turn left
Inflow A	True value (unit)	6,555	1,276	1,583
	Measured value (unit)	6,494	1,244	1,592
	Average of RMSE (unit)	1.5	1.2	1.1
Inflow B	True value (unit)	3,000	1,140	409
	Measured value (unit)	2,934	1,257	428
	Average of RMSE (unit)	1.8	1.9	0.8

with three lanes, and RMSE (root mean square error) of each traffic signal cycle is used as an index showing the measurement accuracy.

In addition, Table 6.3 shows 12-h evaluation results on the radar measurement of direction-specific traffic volume for the inflows A and B in the same rainy day.

In this error analysis, the time zone from 7.00 to 19.00 is chosen, and RMSE is calculated for each measurement data every 5 min. As seen from the average of RMSE, the number of vehicles measured for each direction has the error of 2 units below. Then, the measurement accuracy of millimetre-wave radar has reached 95 per cent or more when evaluated by the total number of vehicles by the directions, except the turn light route of the Inflow B, that is about 90 per cent accuracy. Since this deterioration is mainly related to the coverage near the radar, establishing an angle adjustment method for installation is a remaining issue.

6.6   Conclusion and discussion

As described earlier, the research initiative for applying the latest millimetre-wave radar technology to the roadside sensor at intersections has been carried out. Especially, the clustering technology of the many scattered echoes obtained by the high-resolution radar was implemented and also the roadside sensor system covering the entire of intersection was verified. Now, the 79 GHz radar prototype was installed at the actual intersection, and the technology development is continued to improve the detection accuracy through long-term data acquisition.

Until recently, the millimetre-wave radar has limited scope of application, as it is difficult to visually grasp the raw data as compared to the spreading situation of the visible camera and LiDAR which are the optical systems. However, since the social demand for automation and labour-saving is rising, use cases of millimetre-wave radar will increasing where its features which are robust to the external environment, superior in the speed sensitivity, and compact and thin as the sensor unit could be utilized.

Particularly, as edge computers that process high-resolution sensing data become widespread, various types of sensors are integrated and fused so that more advanced functions can be realized in real time. In the field of road traffic, utilization of the sensing data will be accelerated, because there is expectation for services such as route analysis based on traffic flow measurement and risk prediction by advanced data analysis.

On the other hand, as a communication infrastructure supporting such data acquisition and analysis, the role of wireless communication, which makes it easy to install equipment and update networks, will be more important. Especially, ITS-dedicated radios have been used for information transmission high demand for reliability such as related to safety driving support. Furthermore, in the 5G communication, improvement in real-time performance is achieved with lower latency transmission, so that vehicles and pedestrians are expected to be connected to more robust networks.

And, through these initiatives to evolve the public infrastructure of sensor and wireless communication, we believe that the persistent effort might be necessary to achieve the purposes of avoiding unfortunate traffic accidents and mitigating serious traffic congestion.

Acknowledgements

I would like to thank my colleagues, M. Yasugi, W. J. Liu, and T. Hayashi, for their years of dedication to this work. I am grateful to K. Iwaoka for lending his expertise on traffic engineering. This work is part of the R&D initiative on 'infrastructure radar systems' commissioned by the Ministry of Internal Affairs and Communications as 'next-generation ITS utilizing ICT' for the Cross-ministerial Strategic Innovation Promotion Program (SIP) in Japan.

References

[1] Final Acts WRC-15, World Radiocommunication Conference https://www.itu.int/dms_pub/itu-r/opb/act/R-ACT-WRC.12-2015-PDF-E.pdf.

[2] ARIB STD-T111 Version 1.1 https://www.arib.or.jp/english/html/overview/doc/5-STD-T111v1_1-E1.pdf.

[3] ETSI EN 302 264 V2.2.1 (2017-05) https://www.etsi.org/deliver/etsi_en/302200_302299/302264/02.01.01_60/en_302264v020101p.pdf.

[4] K. Takahashi, H. Yomo, T. Matsuoka, *et al.*, "Evolution of Millimeter-Wave Multi-Antenna Systems in the IoT Era", IEICE Transaction Electronics, 2017.

[5] W. J. Liu, T. Kasahara, M. Yasugi, and Y. Nakagawa, "Pedestrian Recognition Using 79GHz Radars for Intersection Surveillance", IEEE proceeding of the 13th European Radar Conference, 2016.

[6] Cross-ministerial Strategic Innovation Promotion Program for Automated Driving Systems http://en.sip-adus.go.jp/

[7] Y. Nakagawa, Infrastructure Radar System as Next-Generation ITS Utilizing ICT, ITU-AJ New Breeze Vol. 27, No. 3, 2015.

Part III

Traffic state sensing by on board unit

Chapter 7

GNSS-based traffic monitoring

Benjamin Wilson[1]

7.1 Introduction

Global Navigation Satellite System (GNSS) probe-based traffic information services are now commonly used across the world. For the purpose of this chapter, we will use the term 'GNSS' stands for Global Navigation Satellite System which is a standard generic term for satellite navigation systems. GNSS includes all global navigation systems, including GPS, GLONASS, Galileo and other regional systems. GNSS data is highly accurate, and this service provides a scalable delivery platform that will continue to improve as more data become available.

7.2 GNSS probe data

Since the introduction of the smart phone in the mid-2000s, there has been exponential growth in the availability of GNSS probe data to providers to support real-time traffic information. Today's service providers can gather information from smart phones, personal navigation devices, telematics devices (in fleet vehicles) and connected vehicles to create their services.

Besides the obvious benefits of selecting data from high-quality sources, the creation of a data pool based on a heterogeneous fleet is important to deliver a service that is both reliable and flexible.

Volume of providers also provide natural balance that is automatically built into the data pool that consists of both consumer and business fleets, passenger and truck data, probes from regular commutes as well as recreational trips. The inherent diversity of the sources used results in the continuity of probe data all the time, everywhere.

7.3 GNSS probe data attributes

The GNSS probe data collected from all these various devices is passed to the traffic service provider via the mobile network provider and the internet in near real time.

[1]HERE Technology, Melbourne, Australia

Each individual GNSS probe point includes information from the device such as the latitude and longitude, travel direction, travel speed and GNSS time stamp. The service provider uses the aggregated information collected from these devices to create an estimation of the traffic flow speed on each road link.

The freshness of this probe data is critically important to service accuracy. Service providers generally record the GNSS position time stamp of each probe point and use this to prioritise importance of each data input.

7.4   Historical data

Using all the GNSS probe data over the past few years, service providers can build an estimate of the current traffic conditions based on historical norms. This historical estimate provides an understanding of the travel time across all mapped road links down to 5-min intervals per day for each day of week (and holidays). This data is also used as an input to real-time traffic algorithms where limited real-time probe data has been collected.

7.5   Probe data processing

The data processing can be seen as the foundation of service as 'probes' are a lot of data and highly unstructured, and the industry also talks about big data processing. The next process describes the journey from probe data to traffic information on a high level.

Retrieve and Transform → Validate → Map-Match → Filter

First, data is validated from the sources for geo-location, heading, map matching, speed validity and other traffic metrics to filter-flawed data from entering the data stream. Provider-specific formats are converted into a standard internal format, using validations that are appropriate to each individual provider's proprietary format. Next, invalid probe points are filtered out from being used to determine speed values. Third, the location is matched and assigns probes to the roadway links. The GNSS probe points are map matched using sophisticated algorithms. The algorithms have been verified for accuracy using independent ground truth validation. GNSS probes are discarded that do not meet the map matching criteria or have outlier characteristics. During filtering, the map-matched probes are specifically treated to handle high-frequency probes (duplicates) or special situations like tunnels.

Products are fundamentally created from the same set of probe data, certain differences in filtering and processing the data are necessary to ensure that each product is tailored to fit its primary use cases.

Last but not least, all probes are anonymised and data security and privacy regulations are taken into account throughout the complete process.

7.6 Real-time traffic information

Monitoring and also joint-collaboration with partners guarantee the usage of the best probe sources available. Focus is on highly accurate GNSS sources with high reliability minimising large-scale filtering to keep latency low. Such validation for instance guarantees low-latency and freshness of the real-time traffic service. Incoming probe data that does not meet latency requirements (i.e. arrives too long after the registered timestamp) is filtered out during generation of the traffic feeds, since it is no longer useful for providing information about current traffic conditions. However, these probe points are still archived and still constitute part of the dataset used to create historical traffic products (Diagram 7.1).

Probe points are matched against the best map data in real time. Specifically, this means, inaccurate probe points outside of road geometry or unrelated to roads accessible by cars are not filtered. Also, content attributes such as bus/taxi lanes, bridges, points of interests (POIs) such as bus stops or taxi drop-off zones are considered to only use the most relevant probe points to road traffic. Last but not least, granular intersection data, including the availability of traffic lights, plays a significant role to distinguish between red-light phases and heavily congested junctions.

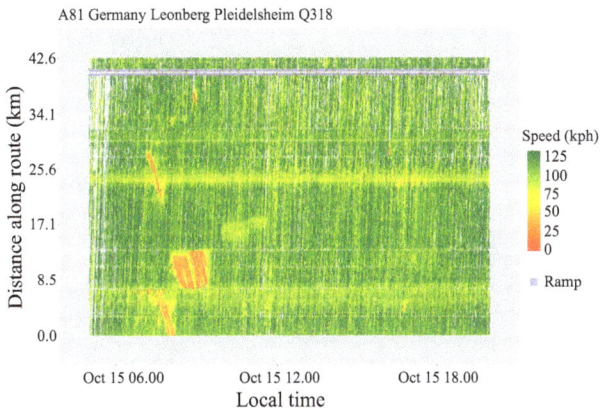

Diagram 7.1   Highly accurate GNSS-based probes

The constant analysis of the probe trace is another important mean to guarantee the best traffic service; analysing the relation between each probe points of a trace, and deciding whether the probe observation is relevant (or not).

7.7   Example of probe data in use

The time–space Diagram 7.1 represents travel speeds collected from GNSS data collected over a section of the A81 in Germany. The colour represents the speed of travel. This represents the raw data that is used to feed calculation of traffic flow speed.

Diagram 7.2 represents the output traffic flow speed created that correlates to the same section road link.

Previously, you can see in the second diagram output speed values represented by colour coding across the various links with the road network.

The previous Diagram 7.2 represents the actually broadcast speed from a smaller link of road represented in Diagram 7.1. The link is between the 8.5 and

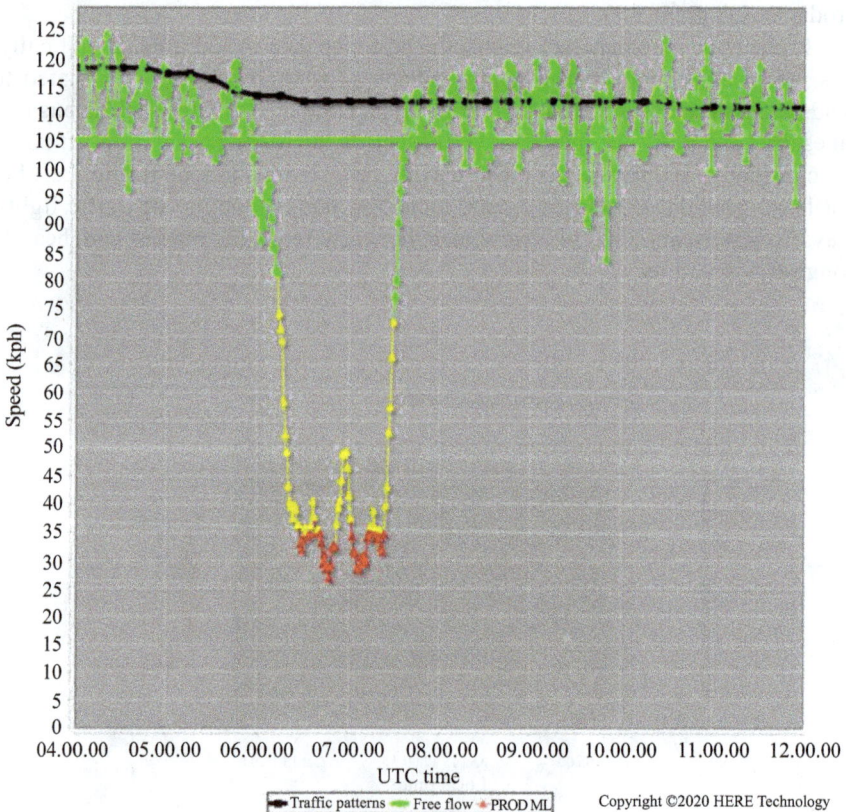

Diagram 7.2

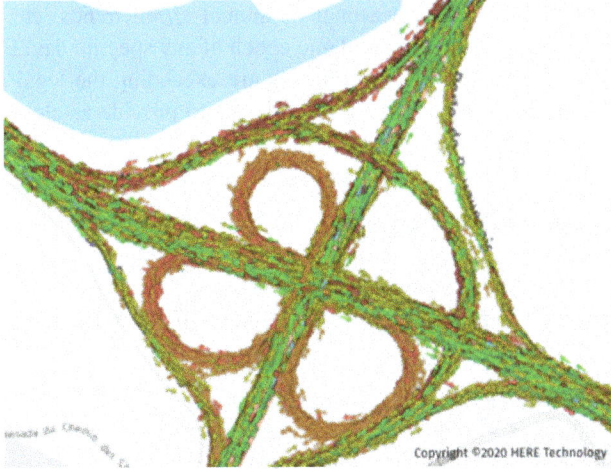

Diagram 7.3

17.1 km section of the road represented in Diagram 7.1. In this road section, congestion appears at approx. 6.00. At this time average, speed reduces from over 100 kph down to below 30 kph. At approx. 7.30, the congestion reduced and speed increase to over 100 kph. Note: the time-scales used between the Diagram 7.1 (local time) and 7.2 (coordinated universal time (UTC)) are different.

7.8   Historical traffic services

7.8.1   Traffic speed average

To enable more realistic journey time estimations based on time and day of travel, traffic speed average data needs to provide average speeds that are not overly influenced by sporadic events (such as accidents, roadworks or adverse weather conditions). As such, this data is aggregated over and up to multi-year periods to eliminate the effect of such unpredictable events. A series of computationally intense processing steps then occur to normalise the data and apply different levels of data modelling depending on the amount of probe data available for each road segment for small time periods (e.g. 15 min). Since this data is primarily intended for routing purposes, outliers are removed, and average speeds are capped to the maximum legally allowed speed limit. The final product is represented by a database of average speeds (e.g. 15-min intervals) across each day of the week (Monday–Sunday) and a holiday average.

7.8.2   Historical traffic analytics information

In contrast, historical traffic analytics information is intended to provide insight into actual average speeds to enable customised 'before and after' analysis, the investigation of daily, weekly, seasonal or annual speed trends, etc. To fit this purpose, data is not aggregated beyond one epoch of any specific day. Furthermore, data is not normalised, or capped if vehicles are exceeding the legal speed limit. No data modelling is employed as the intention is to provide as close to raw data as possible.

7.9   Advanced traffic features

With increase access to large volumes of GNSS probe data, more advanced services are being released to improve vehicle safety and traffic data accuracy.

7.10   Split lane traffic

Split lane traffic provides lane-level precision on highways and ramps, where there is a major discrepancy in speeds between the lanes as opposed to averaging all the lanes together for one averaged speed. With high accuracy mapping data, flow and routing information discrepancies between GNSS data can be attributed to lanes across the road network. This can then be used to better inform drivers where their journey will be impacted by congestion.

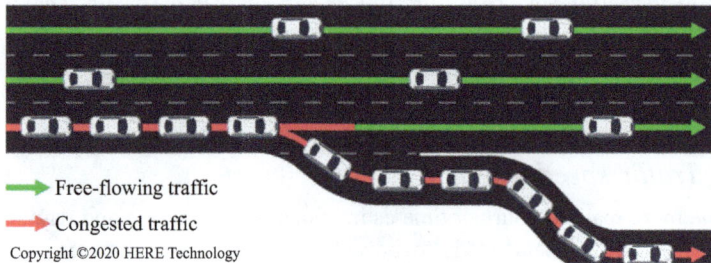

→ Free-flowing traffic
→ Congested traffic
Copyright ©2020 HERE Technology

7.11   Wide moving jam (safety messages)

Research on the movement of vehicle using GNSS data has identified a phenomenon known as 'wide moving jam'. This identifies that congestion moves backwards along the road network. This create a danger of rear-end accidents where vehicles might be travelling at high speeds along the road. Using advance algorithms, this phenomenon can be identified and safety messages sent to vehicles to reduce the risk of accidents.

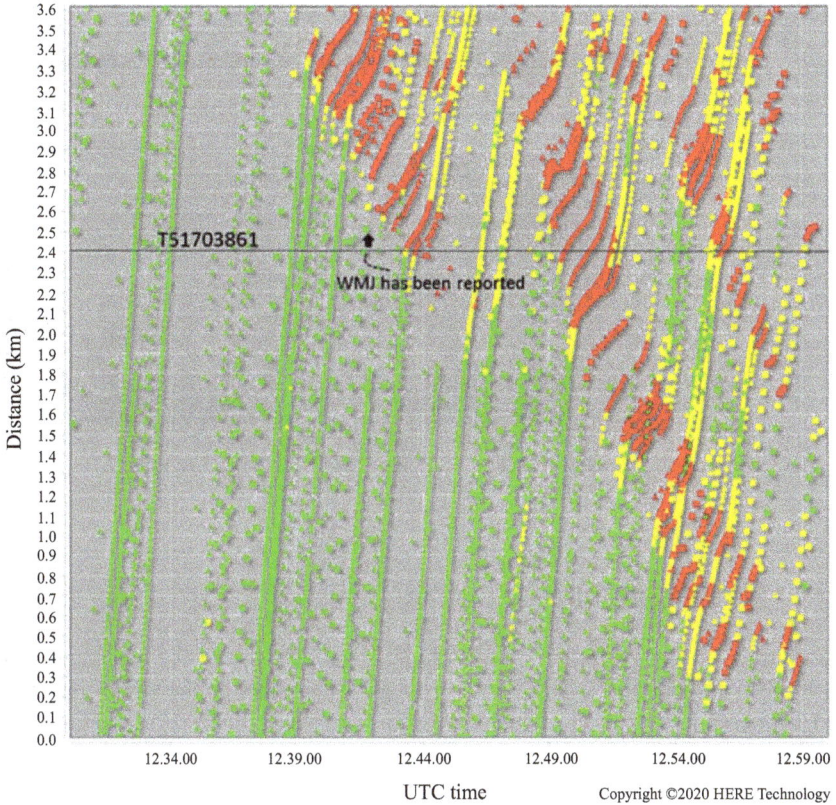

The previous time–space diagram demonstrates the traffic congestion end point moving backwards along the roadway.

7.12 Automated road closures

With the introduction of the machine learning tools combined with GNSS probe data, the verification and detection of road closures can be automated. This enables faster detection of the reopening of closed roads and improves the spatial precision of reported road closures by using GNSS probe data to continuously monitor the road network.

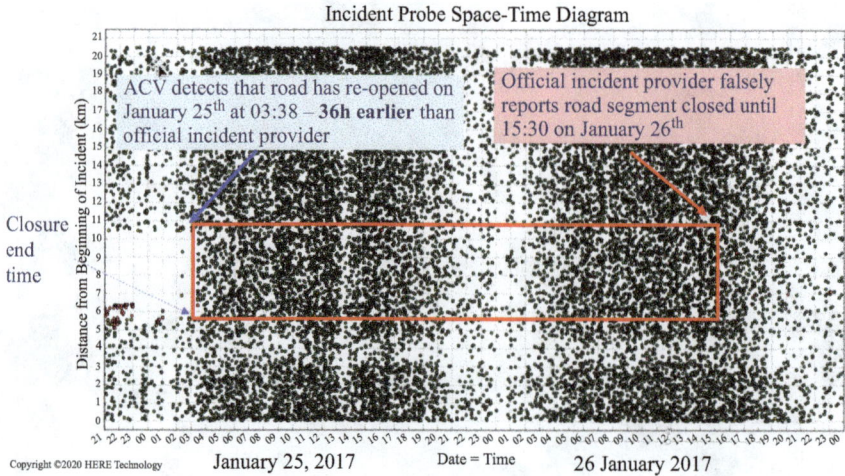

Incident Probe Space-Time Diagram

7.13   Quality testing

There are a variety of testing methodologies available to test the accuracy of the traffic feed and organise which should educate themselves on the various strengths and shortcomings of each testing methodology. Next are three commonly used testing methodologies, although others exist.

7.14   Ground truth testing

A ground truth test entails a vehicle driving in a specific market with a high accurate GNSS tracking device. These tests are generally conducted over a week in morning and evening peak times to ensure capture real market conditions and congestion events. This data is then matched by a testing team against the traffic feed. This provides a highly accurate comparison of real-world conditions to our broadcast feed. The quality results can then be shared in percentage accuracy terms based on the results collected.

7.15   Probes as ground truth

This is an automated testing programme that utilises existing probe data to conduct an automated test of the existing real-time traffic broadcast. This method utilises the probe data collected and then identifies probes that meet certain critical to be used as ground truth. Similar to the previous method, this data is tested against the traffic feed outputs to understand quality. Importantly, automated testing methods such as probes and ground truth prove a 24/7 testing capability for ongoing monitoring of the service.

7.16 Q-Bench

Q-Bench is a traffic-quality measurement method (developed by BMW) and adopted by other OEM's to test quality. The goal of Q-Bench is to create a **standard** method to evaluate traffic quality across different traffic providers. Important to note that Q-Bench creates a standard method from a final metric/measurement perspective only. The Q-Bench method does not prescribe the type of ground truth to be used with it. It is ground truth independent and supports both fields driving and automated testing as ground truth.

7.17 Conclusion

With the continued growth of GNSS/GPS data available as more smart-phone application are release, more vehicles become connected and fleet increasing adoption telematics services, GNSS/GPS-based traffic services will naturally represent a far greater sample of road users. As GNSS/GPS data does not require expensive infrastructure and provides coverage across the entire road network, it has become a scalable and cost-effective method for understand the movements of vehicle across the road network. As more vehicle sensor data become available and continued investment in research, it is expected that more user cases can be serviced by this traffic modelling method.

Chapter 8

Traffic state monitoring by close coupling logic with OBU and cloud applications

Nobuyuki Ozaki[1], Hideki Ueno[1]#, Toshio Sato[2]†,
Yoshihiko Suzuki[2]#, Chihiro Nishikata[3], Hiroshi Sakai[4] and
Yoshikazu Ooba[2]#*

Traffic state monitoring is important for improving and maintaining smooth traffic. Various factors that affect traffic can be considered as states. These factors to be detected can be not only conventional traffic congestion but also the volume of pedestrians at stops or pedestrian flooding into roads and blocking vehicles' paths and bicycles weaving through traffic. The conventional approach to sensing is to deploy roadside units such as loop coil, sonic sensor, or camera sensor. The new approach is based on a vehicle probe system gathering mainly location and speed data. However, a more sophisticated approach is an image-recognition-based probe system. This approach can directly sense traffic states. Various sensing targets can be detected by developing image-processing logic specific to targets in collaboration with cloud applications. In short, the image-recognition-based onboard unit's (OBU's) probe system has flexibility for further possibilities.

With the OBU-based image-processing technique, we have developed the concept and also proved that various traffic states can be estimated by the combination of OBU and cloud applications. Our image–processing large scale integration (LSI) for automotive use, called Visconti[TM], is the key to achieving this architecture. Several pattern recognition tasks can be performed concurrently on this LSI with lower power dissipation. The detection results are sent in plain text to the cloud system from many vehicles so that the system has light bandwidth in the communication channels.

[1]Toshiba Corporation, Infrastructure Systems and Solution Company, Kawasaki, Japan
[2]Toshiba Corporation, Power and Industrial Systems Research and Development Center, Kawasaki, Japan
[3]Toshiba IT and Control Systems Corporation, Tokyo, Japan
[4]Toshiba Social Automation Systems Co., Ltd., Kawasaki, Japan
*Present affiliation: Nagoya University, Nagoya, Japan
†Present affiliation: Waseda University, Tokyo, Japan
#Present affiliation: Toshiba Infrastructure Systems & Solutions Corporation, Kawasaki, Japan

We have conducted two usage case studies with the idea of close coupling estimation with OBU-based image processing and cloud-based applications. In the first usage case, traffic volume is estimated at highways and in the second one, traffic congestion and pedestrian crowds are estimated at arterial roads. Proof of concept for the first usage case was conducted at Tokyo Metropolitan Expressway with an interest group consisting of the Tokyo Metropolitan Expressway and the Tokyo Institute of Technology. The second one was conducted in Melbourne's central business district under AIMES [1], or the Australian Integrated Multimodal EcoSystem, led by the University of Melbourne, in close collaboration with Yarra trams.

8.1   Introduction

Traffic flow rate with density and space mean speed is conventionally sensed by roadside units. Typical approaches of sensing are based on loop coil sensors, sonic sensors, or camera-based sensors. In Japan, these roadside units are placed at roadsides ranging from every several hundred meters to several kilometers [2]. These sensors are reliable as they can count each vehicle passing by. Also, at arterial roads, such as in Tokyo, the number of sensing units at roadsides is most likely more than 20,000. Therefore, traffic management, including signal control, can be based on these reliable sensing devices. One of the disadvantages is the cost to place or build roadside units.

Probe data, from units mounted in vehicles, is assumed to be a cost-effective approach to visualizing traffic congestion [3]. Also, smartphone applications can sense similar data with their own sensing devices. They can provide location data along with vehicle status such as speed. Moreover, various metrics such as traffic patterns, travel time, and travel cost can be calculated using various estimation logic [4]. However, considering the behavior only of one's own vehicle is not enough to understand traffic volume in the area around the vehicle. It may be better to have a larger amount of data, or floating car data, for calculation from smartphone applications [5]. Using existing GPS mounted on buses is a cost-effective way to estimate a network's traffic conditions [6]. Its penetration rate matters when GPS-based probe data is used.

Queue estimation is also important for traffic management as this cannot be sensed directly from existing roadside sensors such as loop coils. Using roadside detectors embedded at enter points, exist points, and inside points of a queue, a queue at an arterial intersection is estimated through simulation based on actual data [7]. With the abundant probe data, queues at highways are also estimated and validated by sending alert messages to stakeholders [8]. Video analysis from the roadside is also another method of detection [9].

Our approach is to mount a camera with an image-processing unit on the vehicle side as an OBU's probe system. The difference from the conventional probe systems mentioned previously is that our approach can sense the surrounding area as a camera is shooting outside of the vehicle and also various locations as it moves. With the various types of image-processing logic to detect objects, this has huge potential for use in sensing traffic states.

We have deployed and evaluated our concept in two usage cases. The first usage case was to estimate traffic volume at highways. Normally traffic volume is estimated by roadside units' sensing speed and number of vehicles to get traffic flow rate, density, and average speed in a certain area. Five vehicles equipped with our camera system traveled along the center circular path of Tokyo Metropolitan Expressway for several rounds. The results obtained by drawing a time–space diagram were quite similar to those of the original data collected by a roadside unit. We have concluded that our approach is effective. The second usage case was to estimate traffic congestion based on queuing at intersections and the volume of pedestrian crowds at arterial roads in the central business district in Melbourne. Our camera system was mounted on a tram. After a 1-year preliminary study by first capturing images from trams beforehand, the algorithms for both OBU image-processing and cloud estimation logic were developed and verified. As the online evaluation, we mounted the camera system into two trams that were in service. We have also concluded that the estimation can be effective in incorporating it into signal control for a more sophisticated method and to improve the service level for passengers riding on trams.

Section 8.2 explains the concept of the "smart transport cloud system." Section 8.3 explains the first usage case and Section 8.4 the second usage case.

8.2   Smart transport cloud system

8.2.1   Concept

Figure 8.1 shows the concept. OBU consists of two units: a stereo camera and a processing unit. This OBU is mounted on public transport vehicles such as buses or trams. After processing image data on board, the detected results along with vehicle information collected through the vehicle network are transmitted through telecom networks to the cloud where information is integrated from each vehicle. The key for this system is the image-processing LSI, Visconti^{TM}, which can process images at OBU and send the detection results to the cloud. Therefore, even if the number of vehicles increases, the telecommunication bandwidth may not be affected seriously as the system sends mainly text data rather than the image itself. There are two reasons to mount the system on public transport vehicles. These vehicles have greater height than normal passenger vehicles, so they have a clearer view for camera units. Another is that they run mainly on main routes where most people care about traffic conditions. In other words, with the data collected, these public transport vehicles can be prioritized over others. The concept was first verified by image-processing logic using a stereo camera on a PC platform [10] and second on an embedded platform [11].

OBU can sense various traffic states by deploying image-processing techniques. Smaller image snapshots can also be transmitted if required. The cloud system gathers information from various public transport vehicles moving around the city and integrates data for traffic state monitoring. The traffic states to be sensed as examples, such as traffic flow, vulnerable road users, and obstacles, are also shown in the figure.

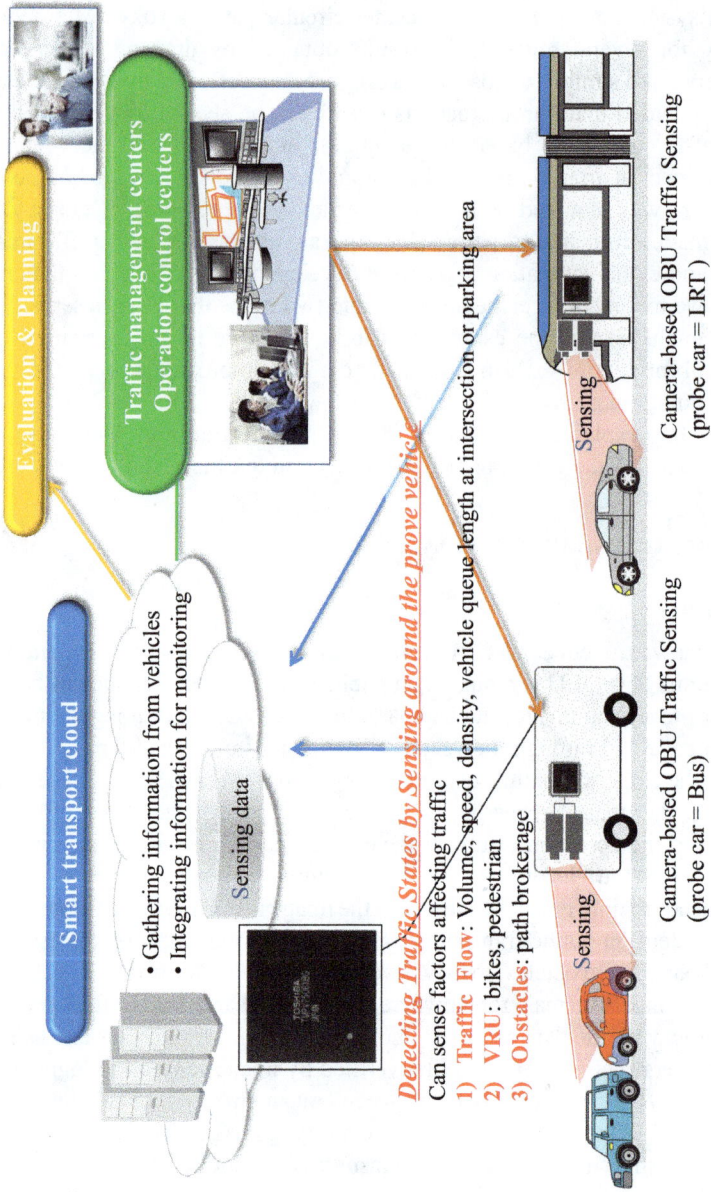

Smart transport cloud

- Gathering information from vehicles
- Integrating information for monitoring

Sensing data

Evaluation & Planning

Traffic management centers
Operation control centers

Detecting Traffic States by Sensing around the prove vehicle

Can sense factors affecting traffic

1) **Traffic Flow**: Volume, speed, density, vehicle queue length at intersection or parking area
2) **VRU**: bikes, pedestrian
3) **Obstacles**: path brokerage

Camera-based OBU Traffic Sensing
(probe car = Bus)

Camera-based OBU Traffic Sensing
(probe car = LRT)

Sensing

Sensing

Figure 8.1   Concept of smart transport cloud

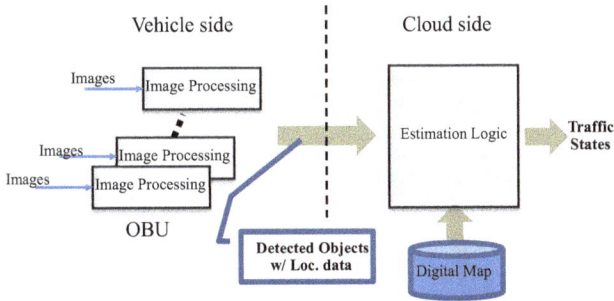

Figure 8.2   Closely coupled applications on two sides

The data can be used in various ways once the data is integrated in the cloud system. The figure shows two examples: one for traffic management and public transport operation control using online data and the other for evaluation and planning using accumulated historical data. Traffic management centers distribute information or control traffic for smoother flow to mitigate congestion. Operation control centers send commands adaptively to public transport vehicles to improve the efficiency of services.

8.2.2   Key technology

8.2.2.1   Tight coupling functions by OBU and server functions

Figure 8.2 shows the concept of tight coupling functions collaborating with OBU image processing and cloud applications in order to estimate traffic states as outputs. Image processing on the vehicle side processes images and sends the results to the cloud, such as pedestrians or vehicles, with location data. The cloud application receives many detected results from different vehicles and estimates traffic states using a digital map.

8.2.2.2   Image-processing technology and its key device

Power dissipation is one of the most concerning features when a vehicle is to be equipped with a unit. Most high-performance LSIs have higher power dissipation as shown in Figure 8.3. Therefore, there is a gap for high-performance LSIs to be used in vehicles. We have overcome this gap by deploying processor architecture and hardware accelerators as shown in Figure 8.4. A media processing engine deploys multicore, very large instruction word, and single instruction multiple data as the architecture. The latest Visconti^{TM} series has eight cores. Typical image-processing logic that consume processing time are embedded into hardware accelerators as shown in Figure 8.4. Crossbar switching is deployed as data bandwidth increases for parallel processing between cores and hardware accelerators.

We have achieved high performance with a low-power-dissipation LSI, called Visconti^{TM}2 and Visconti^{TM}4 [12]. Comparison was made with our own image-processing logic. It takes roughly 4,000 ms on a 1-GHz PC, while on our Visocnti2 platform, it takes only 80 ms, which is 50 times faster than that of PC. Furthermore,

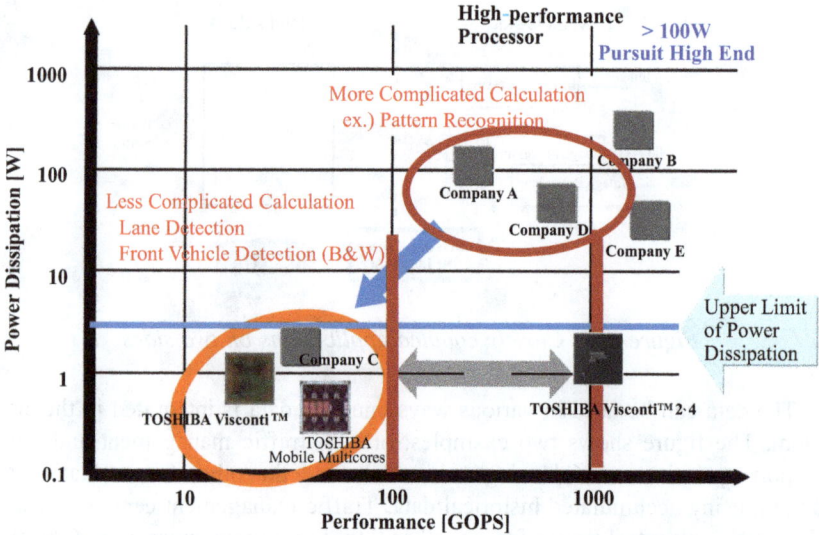

Figure 8.3   Performance vs. power dissipation for embedded system

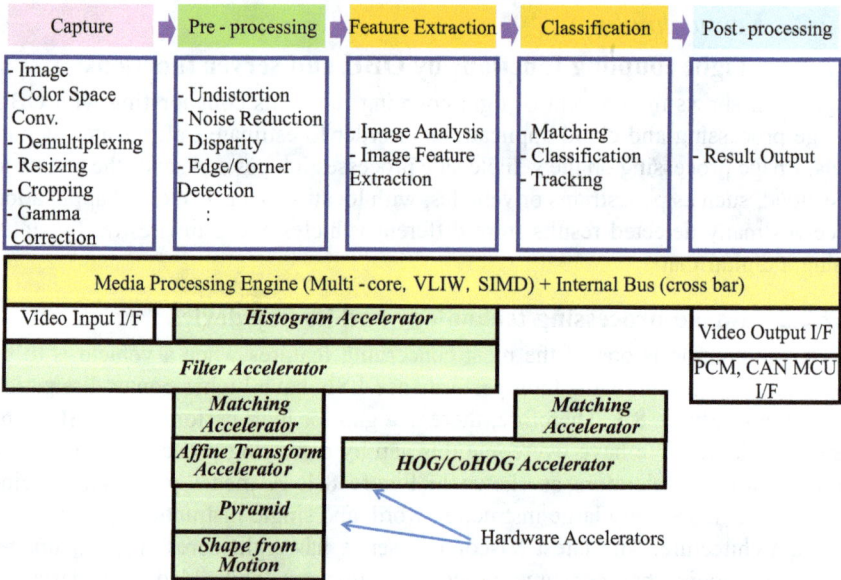

Figure 8.4   Hardware architecture for image processing

Visconti^{TM}4 performs roughly 4 times faster (1900GOPS) than Visconti^{TM}2 with only 1.4 W as the typical condition.

Our approach for image processing is mainly using pattern recognition with the help of a depth map created by stereo camera. Our pattern recognition algorithm is based on CoHOG, or co-occurrence histograms of oriented gradients [13], which has high detection performance, and it is also embedded as one of our accelerators in Visconti^{TM}. Thanks to this high performance, Visconti^{TM} can process several applications concurrently.

8.2.2.3 Cloud applications

Cloud applications estimate traffic states using a digital map after receiving detected objects with location data from each vehicle and integrate and store them to display on graphical user interface (GUIs). States can be traffic volume, traffic congestion, pedestrian or cloud volume alongside roads, or reckless bicycle driving. In order to estimate states, the logic needs to define behavior of detected objects by understanding road structure using GPS location data and a digital map of the area around the vehicle location. Attributes of the road structure information at a certain point can include the number of vehicle lanes, locations of intersections, parking areas, and pedestrian or bicycle pathways. In the case of vehicle detection, behavior can include whether a vehicle is parked legally or illegally, or a vehicle stopping because of a red light at an intersection.

8.3 Usage case 1: estimation of traffic volume at highway

8.3.1 System description

The OBU consists of two units: a 1.2-megapixel stereo camera unit and a processing unit with communication capability as shown in Figure 8.5(a). Two image sensors are synchronized in the camera unit and a combined left and right image is sent to the processing unit. The processing unit uses the Visconti^{TM}2 processor LSI for image

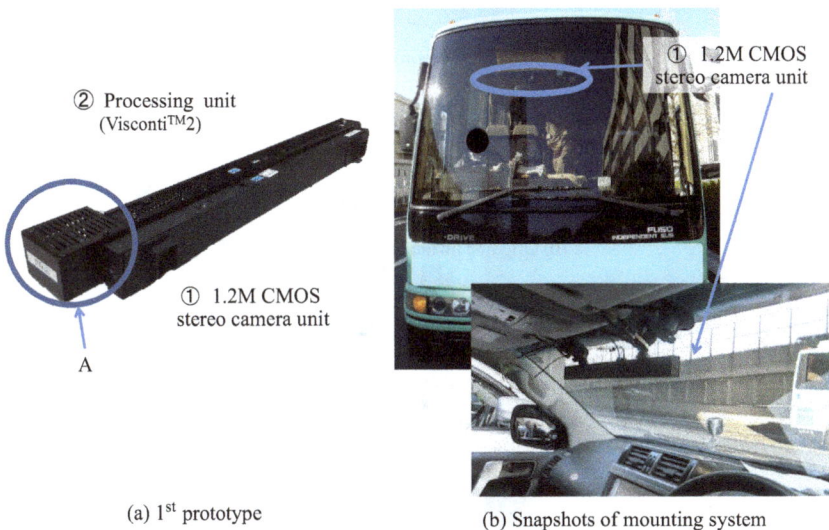

(a) 1st prototype   (b) Snapshots of mounting system

Figure 8.5   (a) First prototype and (b) snapshots of mounting system

processing. The idea behind designing a processing unit is to attach it to the backside of a camera unit in order to combine the units into one. Part A in Figure 8.5 (a) is a peripheral component interconnect (PCI)-express interface device that is available commercially for recording. Recording can be done by connecting a USB3.0-based solid state drive (SSD) unit. This part can be detached if recording is not required. As this was our first trial case mounting our system in vehicles, we focused on recording image data while the vehicles were in operation. Once the data was collected, we analyzed it at our office. Figure 8.5(b) shows an image of the evaluation system mounted in a bus and a sport utility vehicle (SUV). Two buses and three SUVs traveled along a circular path on the Tokyo Metropolitan Expressway.

8.3.1.1   Onboard subsystem

Figure 8.6 shows the step toward creating a time–space domain diagram. During the drive, the processor unit captures combined stereo image from a camera, speed is captured by an in-vehicle network called CAN, and location data is also acquired. Those data are stored on an SSD through a PCI-express connection as shown in Figure 8.6(a).

Image processing is conducted by playing the recorded data from the SSD as shown in Figure 8.6(b) after returning from the drive. By image processing, the distance to the vehicle in front is calculated with the help of lane detection by first detecting the vehicle through CoHOG pattern recognition, and second, finding the detected vehicle's distance using a depth map. The detected results are stored in a temporary file with vehicle speed and location data. The temporary files from all probe vehicles are gathered and fed off-line into a cloud subsystem as detection results to finally estimate traffic state as shown in Figure 8.6(c). Figure 8.7(a) shows the detected vehicles indicated with red vertical lines with the rough distance in red figures.

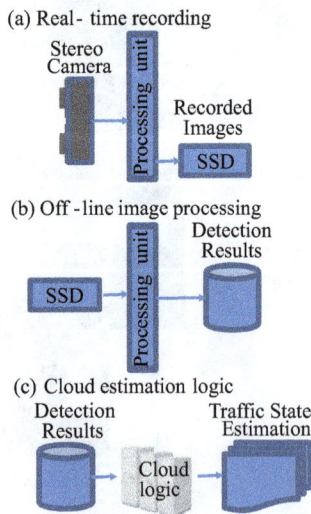

Figure 8.6   Steps toward estimation: (a) real-time recording, (b) off-line image processing, and (c) cloud estimation logic

8.3.1.2 Cloud subsystem with GUI

Figure 8.7(b) shows the GUI for simple traffic monitoring. The cloud system collects detected information from vehicles operating on the highway. The GUI uses colored lines and symbols to display the present traffic situation on a map along with the driving route. The four small blue circles indicate the vehicles' present locations and the three colors along the route indicate the traffic conditions. Congested traffic is indicated by red, smooth traffic by green, and moderate by yellow. This congested condition is defined by vehicle speed as the simple indicator. As probe vehicles move ahead, new colored lines appear with recent past plots remaining behind.

8.3.2 *Traffic volume estimation*

Probe vehicles collect vehicle speed, location data, and the distance to the vehicle in front. When a vehicle is detected in front, its distance is measured from the depth map. Using these three data, the traffic state—flow ($\hat{q}(A)$), density ($\hat{k}(A)$), and speed ($\hat{v}(A)$)—can be estimated as follows, as shown in Figure 8.8 [14]:

$$\hat{q}(A) = \frac{\sum_{n \in P(A)} d_n(A)}{\sum_{n \in P(A)} |a_n(A)|}$$

$$\hat{k}(A) = \frac{\sum_{n \in P(A)} t_n(A)}{\sum_{n \in P(A)} |a_n(A)|}$$

$$\hat{v}(A) = \frac{\sum_{n \in P(A)} d_n(A)}{\sum_{n \in P(A)} |t_n(A)|}$$

where A is the predetermined time–space region (e.g. 5 min × 500 m grid) in the time–space diagram, $P(A)$ is the all probe vehicles in region A, $|a_n(A)|$ is the time–

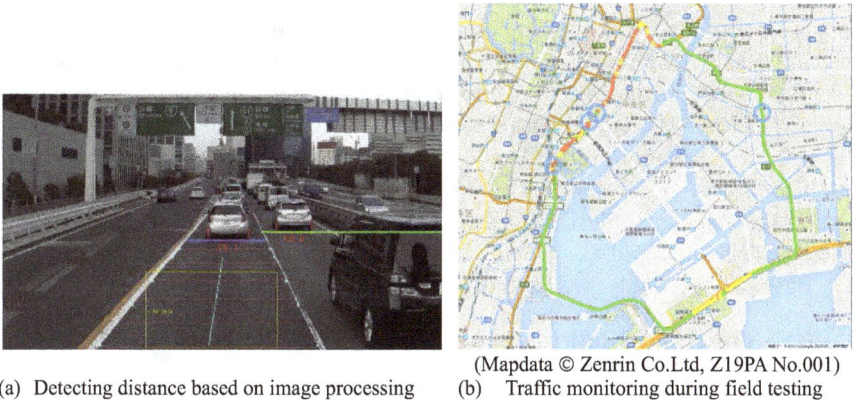

(a) Detecting distance based on image processing

(Mapdata © Zenrin Co.Ltd, Z19PA No.001)
(b) Traffic monitoring during field testing

Figure 8.7 Example of GUI: (a) detecting distance based on image processing and (b) traffic monitoring during field testing

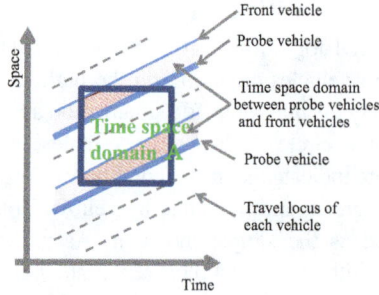

Figure 8.8   Outline of estimation algorithm

space region between vehicle n and the vehicle ahead of it, $d_n(A)$ is the distance traveled by vehicle n, and $t_n(A)$ is the time spent by vehicle n.

The equation can be explained using Figure 8.8 given an arbitrary time–space domain shown as a rectangle in the figure. The probe vehicle measures the distance to the vehicle ahead. It is marked as a red-shaded area. Traffic flow rate, density, and velocity in the specific time domain space can be calculated by the samples driven by probe vehicles.

This approach has the advantage of using distance for estimating the traffic state, and the calculation is conducted on cloud applications.

Figure 8.9 shows the estimated traffic flow rate and density in the time–space diagram calculated with the cloud logic and compared with the reference data. Lines in the second rows show the trajectories of the probe vehicles. Lines of the same color indicate the same vehicles as the vehicles traveling in the circular path of 18 km several times.

The penetration rate of the probe vehicle might be roughly 0.1% when we think about all of the traffic for this circular path. The diagram created by the probe system shows results with a similar tendency compared with the reference data except marked area A. Regarding area A, the estimated version has a higher flow rate and density than the reference. This is because one of our probe vehicles drove very fast so that the distance to the front vehicle was definitely narrower than other probe vehicles. Also looking into the diagram, this vehicle was the only vehicle on the grid for calculation, so the result is skewed by this rough driving. In order to resolve this case, probe vehicles must be driven in accordance with the traffic flow around the vehicle or must increase the penetration rate enough to cover several probe vehicles in a grid as shown in the figure.

Based on this study, we have concluded our approach is effective.

8.3.2.1   Other possibilities for estimating traffic states: vehicle behavior in adjacent lanes

This section explains an example to show further possibilities to make good use of camera shooting of the outside environment.

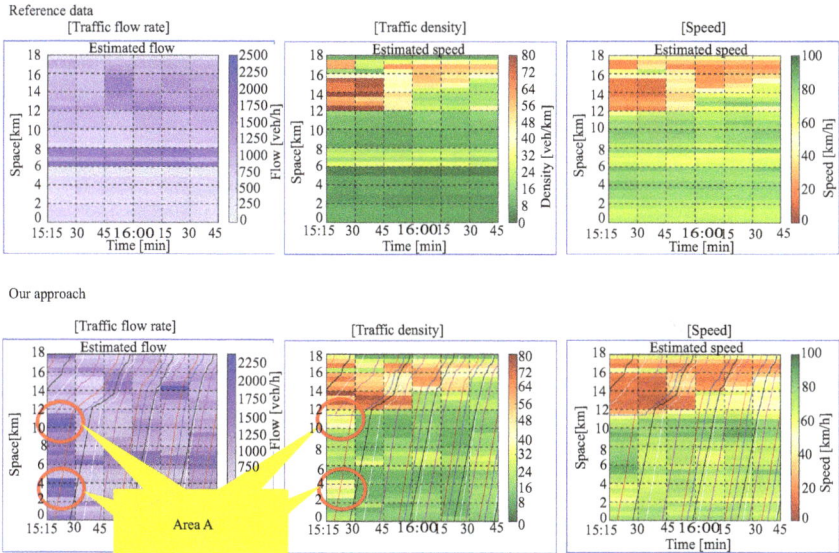

Figure 8.9   Evaluation compared with the reference

Once a vehicle is detected and traced for a moment by image processing for adjacent lanes, the rough vehicle speed can be estimated as the speed relative to one's own vehicle. After plotting the speed of one's own vehicle and vehicles in adjacent lanes, as shown in Figure 8.10, it can monitor different behaviors corresponding to each lane. This is important to monitor vehicle behaviors before merging into a lane or departing from a lane, as traffic is often different at these locations compared to other lanes at the same places. Red marks show that vehicles traveling in the left lane are moving faster than the camera car's lane. This is because the left lane, in which fewer vehicles are driving, is departing from the main path before the junction.

8.4   Usage case 2: estimation of traffic congestion and volume of pedestrian crowds

We carried out proof of concept (POC) with Yarra trams under the AIMES project. POC is verified in several steps: first, defining benefits of how detecting objects can be useful for each stakeholder; next collecting image data to design an image-processing algorithm on OBU and estimation logic in the cloud; and lastly conducting an online demo with a small number of trams.

8.4.1   Benefits from the system

After several discussions with AIMES members, we defined the benefit that the system can provide to several stakeholders shown in Figure 8.11. Stakeholders can

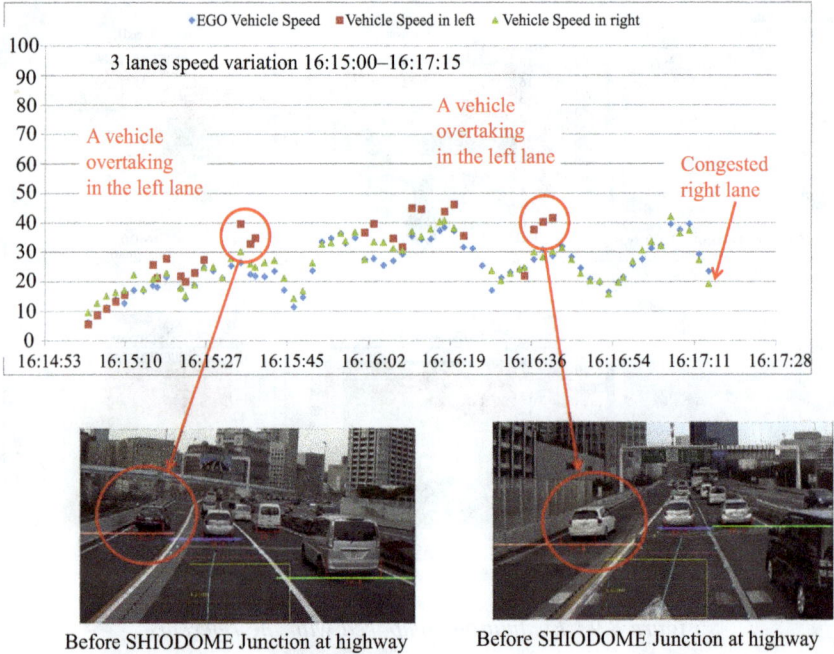

Figure 8.10   Monitoring adjacent lanes

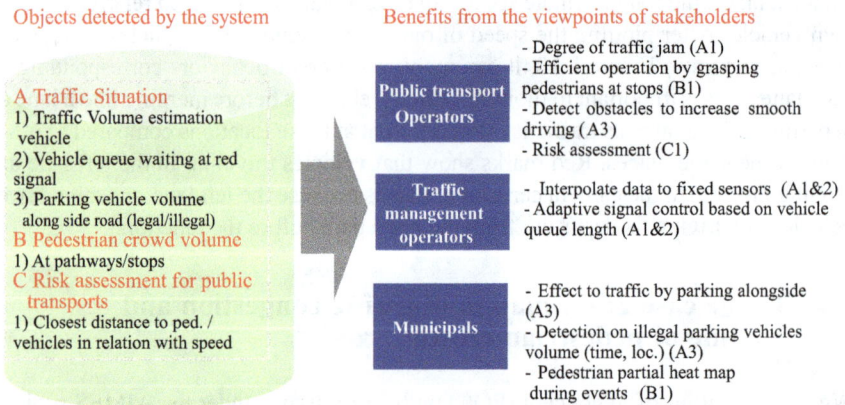

Figure 8.11   Outputs and potential benefits from detected objects

benefit in two ways. The first way is the online information to manage services and traffic better as quickly as possible, and the other is accumulated historical data to analyze and plan for future modifications to infrastructure itself or changes to service patterns. From among these benefits, we selected two cases for POC, as shown in Table 8.1.

Table 8.1   Two cases for POC

Case	Benefits	Objects to be detected	Example
1	Smoother traffic flow by adaptive signal control considering queue length	Vehicle queue lengths waiting at red signal at intersections	
2	Better public transport services by adjusting bumper-to-bumper situation	Pedestrian waiting at tram stops, which caused delay to tram departure time at stops	

8.4.2   System description

8.4.2.1   Architecture for the online study

Figure 8.12 shows the system architecture for the online study. Two trams were in service on Route 96, as shown in Figure 8.13, for roughly a week OBUs equipped on both sides. This Route 96 was selected for the pilot study because it has a wide range of operational characteristics. On various sections of this route, trams operate in traffic where services are affected by queues, partly separated from traffic but where vehicles may be a hazard, in a pedestrian street, and on exclusive tracks. There are also different tram stop designs used on Route 96, with passengers waiting different distances from the tram, and on different sides of the tram, depending on the location. The system is independent from that of the tram system except for the power supply. The detected results are sent through a telecom network to the cloud where estimation logic calculates the length of vehicle queues waiting at intersections and the volume of pedestrians waiting at tram stops.

One year before the online study, we recorded image and data to develop logic. Stereo cameras and processing units were mounted on the upper hatch of the driver's cabin shown, as shown in Figure 8.14(a). The units used were the same as those we used for the highway usage case study, and the steps followed were also

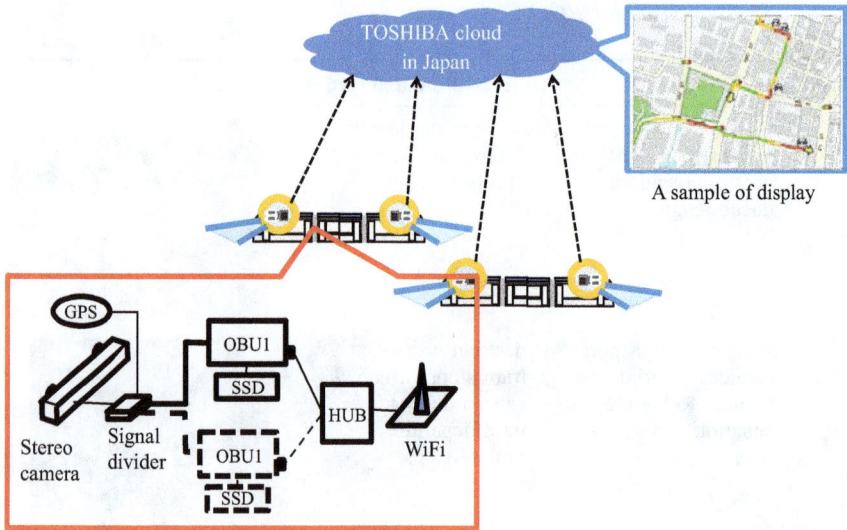

A sample of display

Figure 8.12   Trial system during online study

the same as the highway case. With the recorded data, logic for image processing and related cloud estimation logic were developed and evaluated.

As for the online study, we designed a new OBU using Visconti[TM]4, the next version of LSI, in order to have higher performance. Figure 8.14(b) shows snapshots of the system equipped. The LSI "Visconti[TM]4" is powerful enough to detect two different objects, vehicles and pedestrians, concurrently, through pattern recognition using two different dictionaries. We also recorded captured images during the online study using SSDs for further improvement of our algorithms.

8.4.3   Logic design

Figure 8.15 shows the logic design concept that is tightly coupled with the OBU and cloud applications. The OBU side detects blocks of vehicles with the following items: rough length of the block and its nearest block expressed as lateral and longitudinal distance from the tram. Concurrently it detects the rough volume of pedestrians at stops, with the distance.

Vehicle and pedestrian detection are conducted through hybrid logic deploying CoHOG-based pattern recognition and a depth map. A pattern-recognition-based algorithm is sometimes not enough, as there might be some partial occlusion by objects, mainly vehicles and pedestrians, so the depth map is also used to find partially occluded objects. Every 300 msec, the OBU performs image processing to detect seven items as shown in Figure 8.15. Results are sent to the cloud side every 2 seconds after packing several 300 msec block data during the 2 second interval.

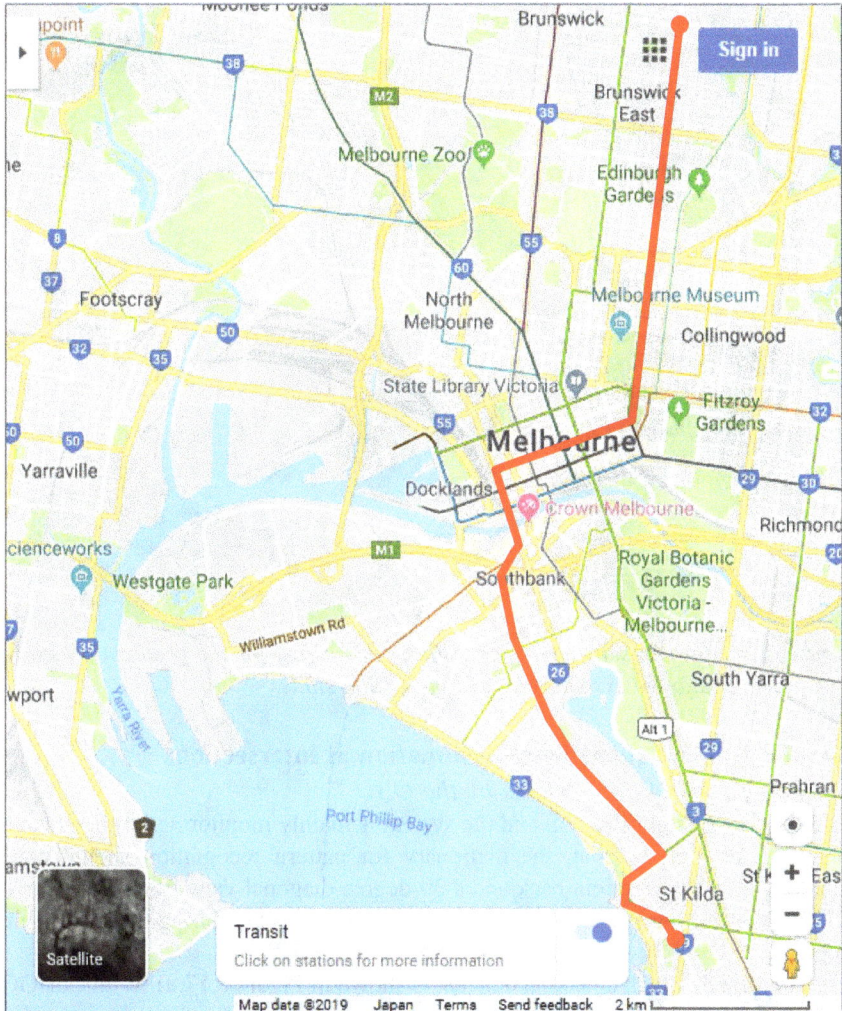

Cited from Google Map

Figure 8.13   Route 96

Cloud applications receive the data from two trams. Both ends from a tram total four OBUs. GPS location data needs to be matched with the route map when the cloud receives data from the vehicle side, as GPS data fluctuates, especially when moving through a city canyon area, even receiving the signal from the quasi-zenith satellite system "Michibiki." The location with peak fluctuation is where tall buildings are facing the road. Fluctuation is shown in Figure 8.16. As a tram runs in both directions, the lead end needs to be determined by the two ends' GPS data sampling from two different times. Once the lead end is determined, the logic needs to understand current road structure.

(a) System setup for the preliminary study

(b) System setup for the online study

*Figure 8.14   System setups for trams: (a) system setup for the preliminary study
and (b) system setup for the online study*

8.4.3.1   Vehicle queue length estimation at intersections

Detection of blocks of vehicles by the OBU

As the camera height is 2.5 m and the system is mainly monitoring left hand lanes, including straight in front, the dictionary for pattern recognition should cover vehicle views from straight back and a 30-degree diagonal view from the height of 2.5 m. The learning set data is created by the data collected during the preliminary studies.

The vehicle block detection concept is shown in Figure 8.17(a). With CoHOG-based pattern recognition, vehicles, the almost whole backside of which can be seen from the system, can be detected. Vehicles occluded by other vehicles are selected by depth map by finding similar height near the detected vehicles. Figure 8.17(b) shows an example of screenshots with the detection markings: red rectangles indicate vehicles detected by CoHOG and green areas indicate the selected area by the depth map.

Vehicle queue length estimation by cloud applications

After receiving detected blocks of vehicles, it checks if each block is moving or not by tracking the same detected vehicles in the block for several frames.

If a block of vehicles is stationary, it marks the near and far end location of the block on a map as it decides it is stationary. As a tram moves, a new area is detected and it is determined if it is stationary or not. If stationary, the length will be added to the previous length. These blocks of length become the candidates for vehicle

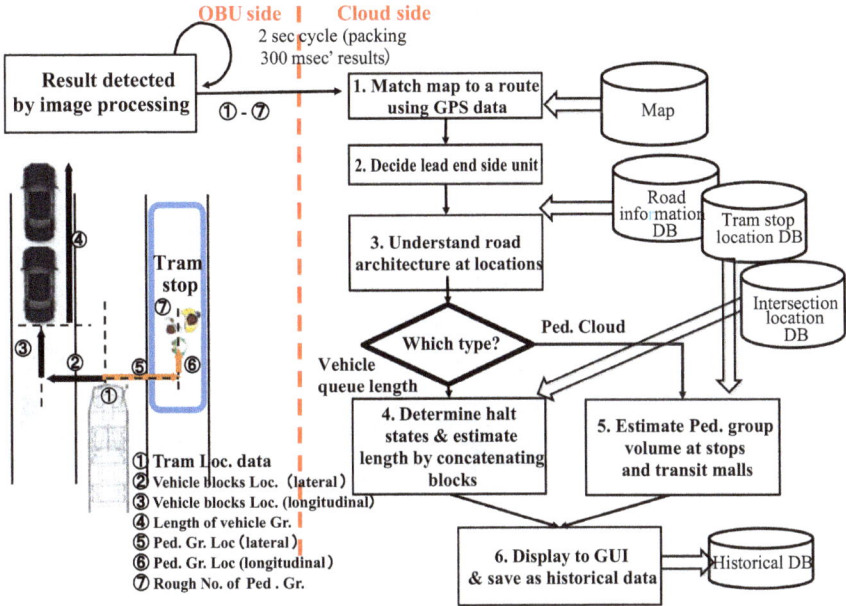

Figure 8.15   Logic concept combining OBU and cloud side

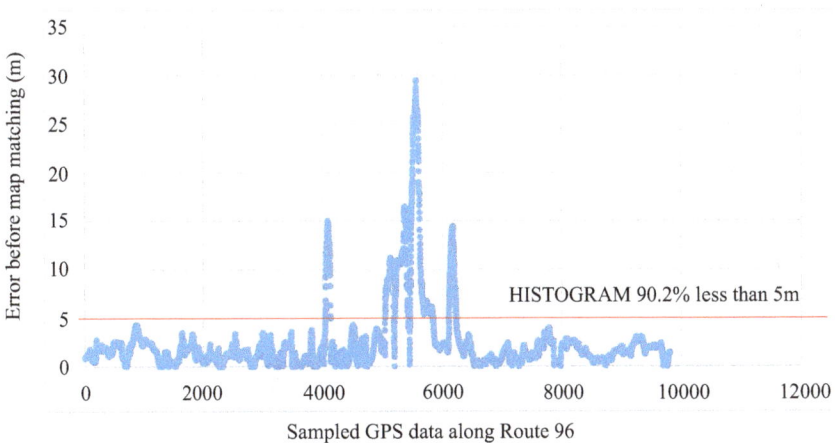

Sampled GPS data along Route 96

Figure 8.16   Fluctuation of GPS location data

queue length at the intersection. Once a tram comes to the intersection and blocks are still stationary, the total accumulated length becomes the final length of the queue waiting at the red light. We also introduce the concept of reliability to the queue length drawing by either elapsed time after detection or distance to the intersection as shown in Figure 8.18. The more time elapses, reliability decreases,

(a) Concept of logic   (b) Detection result

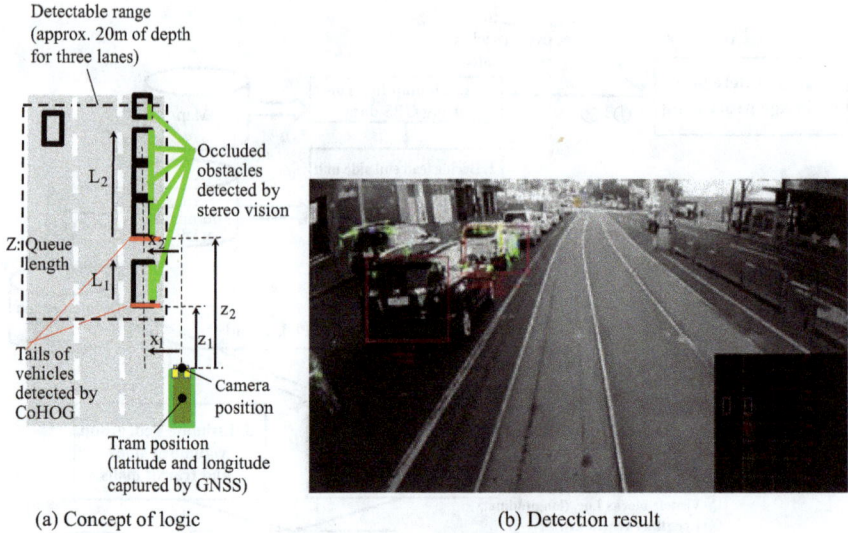

Figure 8.17   *Concept of logic to detect blocks of vehicles: (a) concept of logic and (b) detection result*

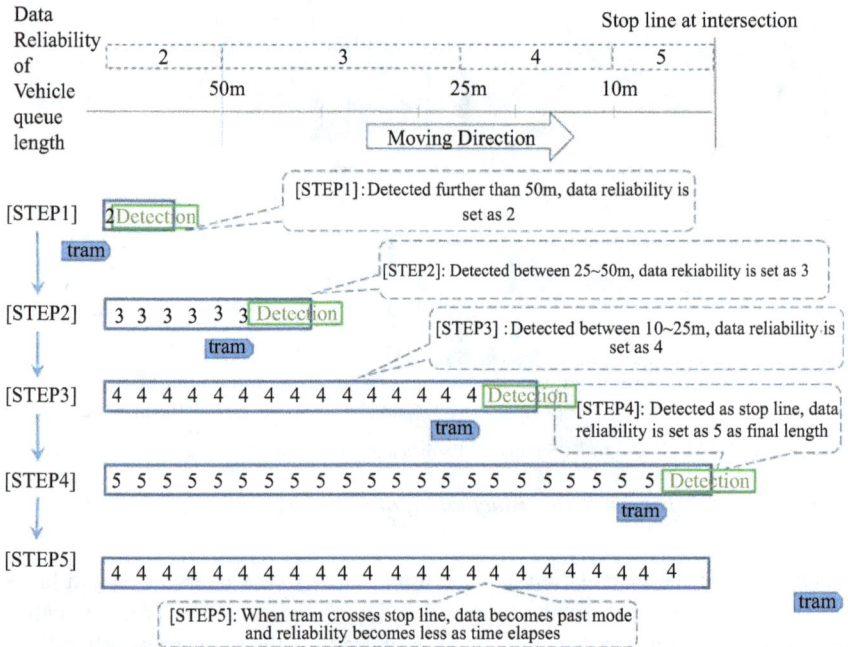

Figure 8.18   *Concept of logic to estimate queue length*

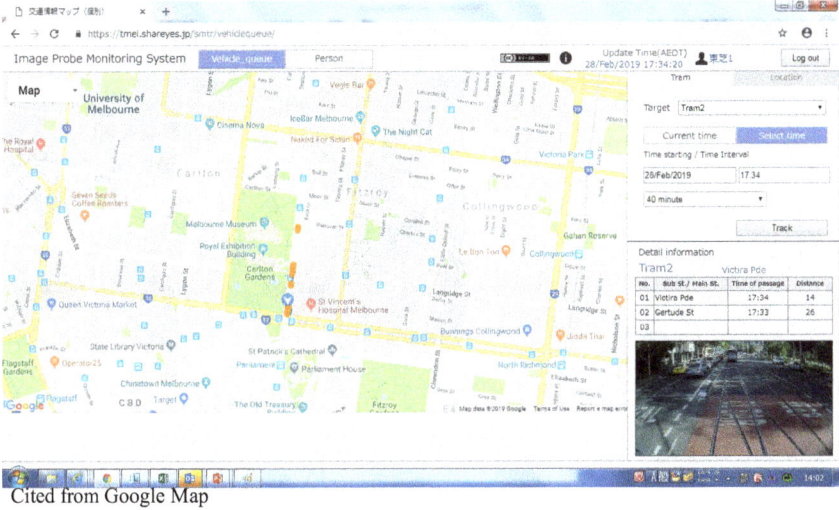

Cited from Google Map

Figure 8.19 GUI for vehicle queue length estimation

and the shorter the distance, reliability increases. Less reliability is shown on the map with a light, diminishing color.

A GUI showing vehicle queue length is shown in Figure 8.19. The arrow indicates the position of a tram. The orange lines indicate queues where vehicles are stopped. The table in the right-hand area indicates the length of queues at previous intersections.

8.4.3.2 Estimation of pedestrian volume at tram stops

Detection of pedestrian blocks with a rough number

Figure 8.20(a) explains the logic concept. Pedestrians are first detected through CoHOG-based pattern recognition. Pedestrians that can be found are mainly those standing at a certain place where a camera can see most of them. Pedestrians occluded by detected pedestrians in front of them are selected based on having a location similar to that of the detected pedestrian extracted from the depth map. Figure 8.20(b) shows an example of a snapshot of detection. The red rectangle is detected through CoHOG and the red-shaded area is selected by a depth map.

Pedestrian volume estimation

If the detected pedestrian blocks are in the area of a tram stop, the location is marked on a map with a circular symbol. Larger symbols indicate larger volumes of pedestrians.

Figure 8.21 shows the GUI. Orange circles indicate the locations of pedestrians and larger circles indicate a larger volume of pedestrians. The table to the right indicates the degree of pedestrian volume at previous tram stops, normalized roughly by stop area size. If a stop is crowded, it shows in red, followed by yellow and green, indicating fewer pedestrians at a stop.

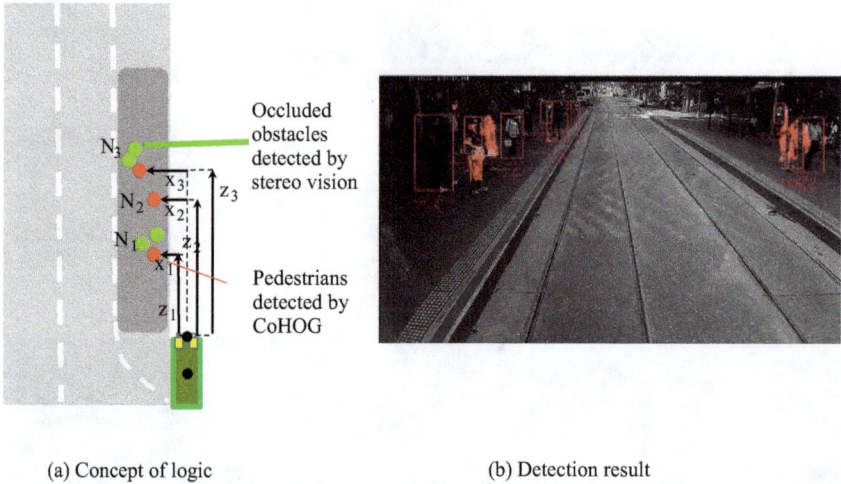

(a) Concept of logic

(b) Detection result

Figure 8.20 Concept of logic to detect pedestrian cloud: (a) concept of logic and (b) detection result

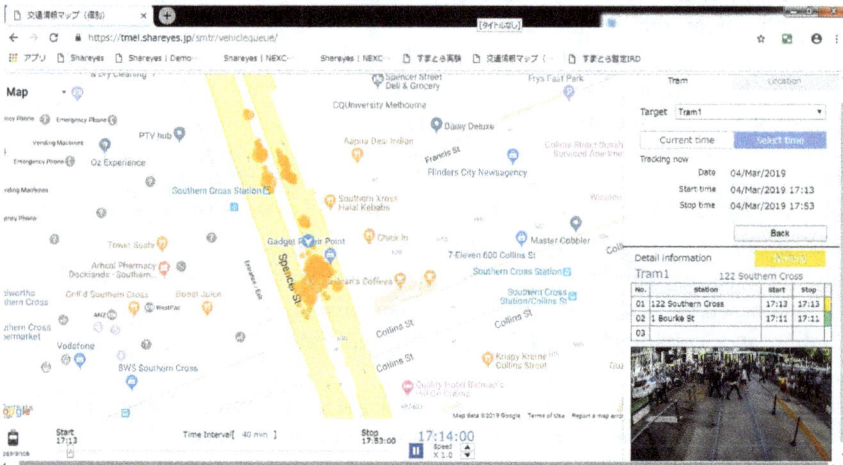

Cited from Google Map

Figure 8.21 GUI for pedestrian cloud volume estimation

8.4.4 Evaluation

Estimated queue lengths were evaluated with a reference in which we manually defined length by checking video. This was conducted before the online study using the preliminary data. Figure 8.22(a) shows the error evaluation compared with the manually measured length and the error length. Roughly speaking, the error for shorter

(a) Estimation of vehicle queue length

(b) Estimation of Ped. cloud volume

Mean of absolute diff (No Ped.)	Pattern Reg	Pattern Reg +Stereo
	2.089	1.964

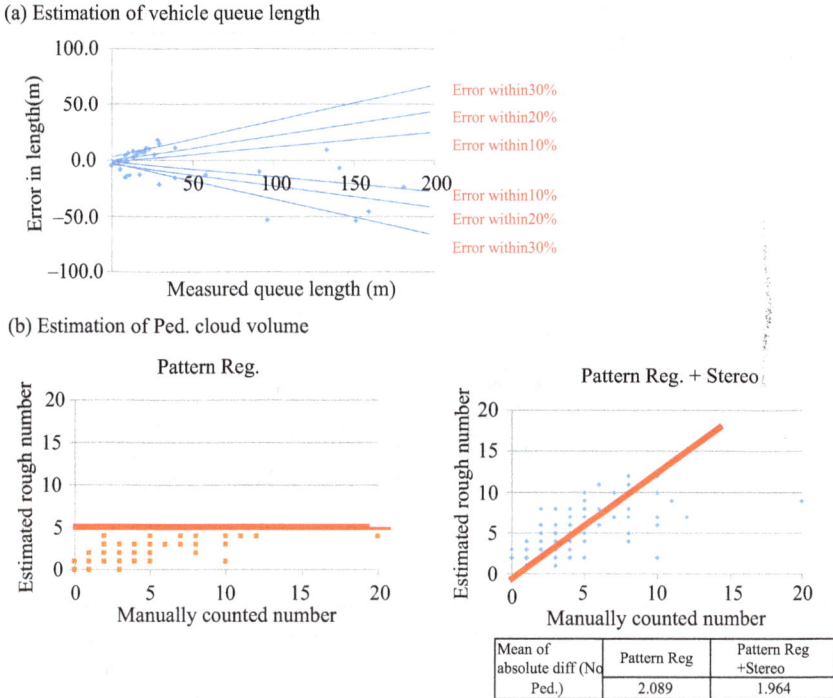

Figure 8.22 Evaluation: (a) estimation of vehicle queue length and (b) estimation of pedestrian cloud volume

queues is within 30% and for longer ones is mostly within 20%. One of the major errors in length may come from GPS longitudinal errors because there is no way to correct data as it is anyway on the road. Also, we understand that we need to improve image-processing logic as much as possible by collecting various situations.

In order to evaluate the estimate of pedestrian cloud volume, we compared it with numbers counted manually by watching video sequences. The algorithm combined with stereo has a better detection rate than just using pattern recognition shown in Figure 8.22(b). However, if the number becomes larger, such as 20 people, it is also difficult to count manually as many are occluded with others packed in smaller areas. Knowing this in advance, we rate pedestrian cloud volume at several levels, such as small, medium, and large.

8.4.5 Other possibilities for estimating traffic: finding parked vehicles

The system has the ability to monitor the far-left side of the road, which may be a parking area, to find whether vehicles are parking legally or illegally. Figure 8.23 shows an example of detecting parked vehicles. The detection is also done with the combination of the OBU and cloud applications. Once road structure has detailed

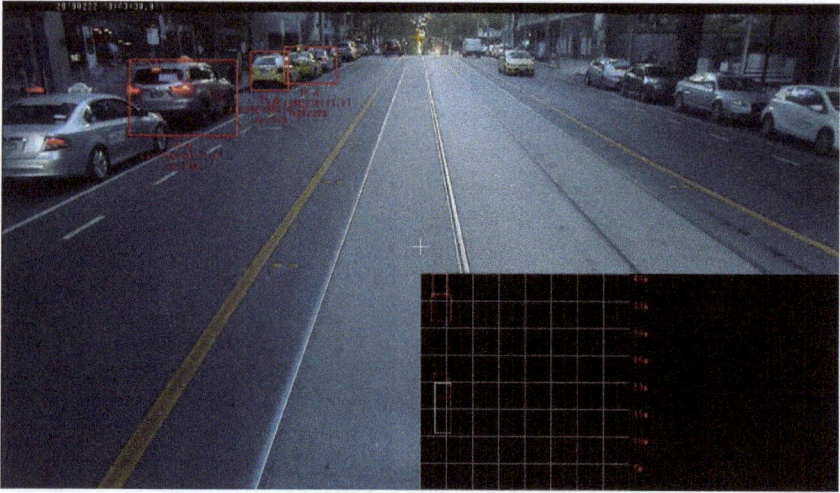

Figure 8.23   Possibilities to detect parking vehicles

information, such as locations for parking areas, with time periods during which vehicles can park, cloud logic can determine whether stationary vehicles are in the area of parking zones or not. If they are not in parking zones during allowed times, the system can report the location and the vehicle's snapshot to law enforcement.

8.5   Conclusion

The concept of the smart transport cloud system has been explained. The system makes good use of high performance with a low-dissipation LSI on the vehicle side to detect first some candidates to estimate traffic states. After the results arrive on the cloud side, traffic states are estimated by cloud logic using the candidates. Processing images on the vehicle side has many advantages, such as the lower bandwidth used for uploading data even if the number of vehicles increases. OBU and cloud applications are closely coupled to estimate traffic states. We have conducted two usage case studies as proof of concept with the help of stakeholders to determine whether our concept is effective. In the first usage case, we estimated traffic volume: traffic flow rate, density, and speed, on the highway. This was conducted and the concept was verified on the Tokyo Metropolitan Expressway. In the second usage case, we estimated vehicle queue length waiting at red lights at intersections and roughly estimated the volume of pedestrians waiting at the stop. Information regarding vehicle queue length can be fed into signal control for more sophisticated traffic control. Information regarding the rough pedestrian volume can be used by tram operators to improve service quality.

An image-recognition-based probe system has huge potential for the development of smart mobility.

Acknowledgments

We are very grateful to everyone who worked with us or gave us pieces of advice during two usage case studies. As for usage case 1, we would like to thank Dr. Hiroshi Warita and Mr. Masayuki Matsushita from Metropolitan Expressway Co. Ltd. and Dr. Toru Seo, Dr. Takahiko Kusakabe, and Prof. Yasuo Asakura from the Tokyo Institute of Technology for their valuable input, including providing reference data from Metropolitan Express Co. Ltd., regarding our project. As for usage case 2, we would like to thank Mr. Ian Hopkins and Mr. Xavier Leal from Yarra trams and Prof. Majid Sarvi of the University of Melbourne for their valuable input as well.

We would not have been able to achieve these results without their help and advice.

References

[1] M. Sarvi, *Australian Integrated Multimodal EcoSystem (AIMES)*, Proceedings, 25th ITS World Congress, Copenhagen, 2018.
[2] T. Murano, T. Watanabe, A. Murakami, and H. Taguchi, *New "System05" traffic control system for Tokyo metropolitan expressway*, Toshiba Review, Vol. 61 No. 8 (2006), pp. 16–19.
[3] T. Oda, M. Koga, and S. Niikura, *A Probe Data-Based Approach of Enhancing Traffic Flow Observation for Traffic Management Applications*, Proceedings, ITS Word Congress, Tokyo, 2013.
[4] M. Sandim, R. J. F. Rossetti, D. C. Moura, Z. Kokkinogenis, and T. R. P. M. Rúbio, *Using GPS-Based AVL Data to Calculate and Predict Traffic Network Performance Metrics: A Systematic Review*, 2016 IEEE 19th International Conference on Intelligent Transportation Systems (ITSC), Rio de Janeiro, Brazil, pp. 1692–1699, 2016.
[5] V. L. Knoop, P. B. C. van Erp, L. Leclercq, and S. P. Hoogendoorn, *Empirical MFDs Using Google Traffic Data*, 2018 IEEE International Conference on Intelligent Transportation Systems (ITSC), Maui, Hawaii, pp. 3831–3839, 2018.
[6] H. Kim and G. Chang, *Monitoring the Spatial and Temporal Evolution of a Network's Traffic Condition With a Bus-GPS Based Pseudo Detection System*, IEEE International Conference on Intelligent Transportation Systems (ITSC), Maui, Hawaii, pp. 3091–3096, 2018.
[7] A. Amini, R. Pedarsani, A. Skabardonis, and P. Varaiya, *Queue-Length Estimation Using Real-Time Traffic Data*, 2016 IEEE 19th International Conference on Intelligent Transportation Systems (ITSC), Rio de Janeiro, Brazil, pp. 1476–1481, 2016.
[8] M. Mekker, H. Li, J. McGregor, M. Kachler, and D. Bullock, *Implementation of a Real-Time Data Driven System to Provide Queue Alerts*

to Stakeholders, 2017 IEEE 20th International Conference on Intelligent Transportation Systems (ITSC), Tokyo, Japan, pp. 354–359, 2017.

[9]   D. Ma, X. Luo, S. Jin, W. Guo, and D. Wang, *Estimating maximum queue length for traffic lane groups using travel times from video-imaging data*, IEEE Intelligent Transportation Systems Magazine, Vol. 10 No. 3 (2018), pp. 123–134.

[10]   K. Yokoi, Y. Suzuki, T. Sato, T. Abe, H. Toda, and N. Ozaki, *A Camera-Based Probe Car System for Traffic Condition Estimation*, Proceedings, 20th ITS World Congress, Tokyo, Japan, 2013.

[11]   N. Ozaki, H. Ueno, T. Sato, *et al.*, *Image Recognition Based OBU Probe System for Traffic Monitoring*, Proceedings, 22nd ITS World Congress, Bordeaux, France, 2015.

[12]   Y. Tanabe, M. Sumiyoshi, M. Nishiyama, *et al.*, *A 464GOPS 620GOPS/W Heterogeneous Multi-Core SoC for Image-Recognition Applications*, Proceedings 2012 IEEE International Solid-State Circuits Conference, San Francisco, USA, 2012.

[13]   T. Watanabe, S. Ito, and K. Yokoi, *Co-Occurrence Histograms of Oriented Gradients*, The Third Pacific-Rim Symposium on Image and Video Technology (PSIVT2009), Tokyo, Japan, January, pp. 13–16, January 2009.

[14]   T. Seo, T. Kusakabe, and Y. Asakura, *Estimation of flow and density using probe vehicles with spacing measurement equipment*, Transportation Research Part C: Emerging Technologies, Vol. 53 (2015), pp. 134–150.

Part IV

Detection and counting of vulnerable road users

Chapter 9

Monitoring cycle traffic: detection and counting methods and analytical issues

John Parkin[1], Paul Jackson[2] and Andy Cope[3]

This chapter considers the importance of cycle counting and monitoring and the nature of cycle traffic. It provides an overview of current methods, including surface, subsurface and above-ground detection methods. It discusses some of the analytical issues in connection with the data and provides a forward view into the future of cycle counting.

9.1 Introduction

9.1.1 *Importance of monitoring cycle traffic*

Cycling is increasingly recognised as an important mode for improving accessibility that also assists in enhancing the efficiency of total traffic movements in urban areas. Some north European countries, the Netherlands and Denmark in particular, have high proportions of cycle use of up to nearly 50% of all trips in some cities. Other countries and cities have much lower volumes, but some cities are seeing rapidly increasing volumes of cycle traffic. With a base year of 2000, London has seen a 136% increase in cycling trips in the period to 2016 [1], and Seville has seen an increase of 535% to 2013 [2]. While the growth rates are dramatic in these leading cities, other places are either seeing less growth or perhaps only marginal change. Many places are, however, planning for growth in cycling.

Data on cycling levels is needed for broadly three purposes. The first purpose is to feed data into aggregated statistics about transport use at a city-wide level. This might then allow for estimates of mode share, for example. The second purpose is for designing and appraising cycling schemes. Data on existing traffic volumes and flow patterns will help in identifying the most suitable new routes for cycle traffic, and also in the estimation of the benefits of a cycle traffic scheme.

[1]Centre for Transport and Society, University of the West of England, Bristol, United Kingdom
[2]Tracsis, Leeds, United Kingdom
[3]Sustrans, Bristol, United Kingdom

Table 9.1 Summary of cycle scheme outcomes, metrics and monitoring tools

Typical outcomes	Metrics	Monitoring tools and data sources
Increased volume of cycling	Change in the number of cycling trips Change in the number of people cycling	Count data User surveys and count data
Increased volume of cycling for specific trips (e.g. to schools and workplaces)	Change in number of cycle trips to a destination Change in the number of people in surveys stating they cycled to the destination	Count data and/or counts of parked cycles Destination-based travel surveys
Increased levels of cycling in specific demographics	Change in the proportion cycling by gender Change in the proportion cycling by other specified socio-demographic group	User surveys and classified count data User surveys and classified count data
Improved actual and perceived accessibility to local amenities and transport hubs	Change in number of bicycles parked at destinations and transport hubs The number of people in surveys who indicate a change in the level of accessibility	Counts of parked cycles User surveys
Improved actual and perceived cyclist safety	Change in the number of cycling collisions and injuries recorded Change in the number of people in surveys who agree that safety has improved	Collision record data User surveys

The final purpose is to evaluate a scheme after it has been implemented [3,4]. Such evaluations require knowledge in four dimensions: the resource *inputs*; the *outputs*, for example in terms of infrastructure constructed; the *outcomes* in terms of the changes in levels of use and finally the *impacts*, for example in relation to changes in levels of carbon emissions through reduced motor vehicle kilometres or improved health and well-being. The findings from evaluations are also important as inputs for subsequent transport planning, such as policy development, future demand estimation and future scheme scoping. They can also provide evidence about the parameters to use for economic appraisal.

Table 9.1 summarises typical scheme outcomes, some of which are supported by cycle monitoring and counting. The table also shows the metrics and monitoring tools and data sources.

Count data can be of various types and simply include volume or be classified in some way, for example by gender of rider. Counts may be undertaken by a surveyor or be collected using detection and recording equipment. User surveys may be of four broad types as follows: within the household, intercept surveys within the journey, based at a destination such as a workplace or school and on-line survey.

9.1.2   Nature of cycle traffic

One of the beneficial features of a cycle is that it is smaller and more efficient than a motor car as a result of causing less delay to other road users. However, this means that it is in fact harder to detect and therefore count when in motion within the transport network. Cycles are heterogeneous and include bicycles, tandems, tricycles, hand-cranked cycles, cargo cycles, cycles adapted to carry children and cycles adapted for use by disabled people [5].

In addition to the variable nature of the cycle itself, the routes that cyclists can use are generally more extensive than routes for motor traffic. Cyclists can use all of the following: the public highway, cycleways, cycleways shared with pedestrians, other public rights of way such as bridleways and canal towpaths and permissive routes over private land. Even when cyclists are catered for within the public highway, they may be located in different places within the cross-section as follows: on the carriageway (with or without a marked cycle lane); on the carriageway and provided with some form of segregation (sometimes called 'light' segregation); on stepped cycle tracks (kerb separated from both the adjacent carriageway and footway) and on fully segregated cycle tracks. Although not regarded as good practice [5], provision may also often still be shared with pedestrians on the footway.

Motor traffic volumes vary by hour of the day, by day of the week and by month of the year. Cycle volumes also vary, but often in a much more pronounced way. Within day, variation may be more pronounced as a result of the cycle being more predominantly used in some locations just for commuting, for example in city centres. Greater variation by day of week may result from a bias towards use on a certain day for certain journey purposes, for example leisure rides on a weekend. Variation by month of year may be pronounced as a result of climate and the number of daylight hours. Data for Greater Manchester, for example, demonstrates this enhanced variability [6]. This greater variation has implications for the timing and duration over which to take cycle movement observations.

Cycling can also be a social activity and cycling two abreast is legally allowed and can be common (even if it is not currently common in some countries with low levels of cycle use). When cycle flows are large, cyclists can and do move with each other, and around each other, in a manner which is fluid in nature. This is in contrast to motor traffic flow, which is observed as, and modelled as, the flow of discrete objects within a system which follow each other.

The variabilities in the nature of the cycle, the nature of provision for cycle traffic and the nature of the patterns of flow of cycle traffic all lead to special challenges in detecting cycles and counting them. This chapter deals with these issues. Section 9.2 reviews current technology and methods for detection and counting. Section 9.3 evaluates the methods in terms of their accuracy and appropriateness for use in analysis. Section 9.4 considers innovations.

9.2   Current methods of detecting and counting

9.2.1   Overview

Count data can be used to determine the number of cycles on a specific route at a specific location. Individual cycle counts can also be aggregated with other cycle counts to determine the total volume of cycle traffic entering and leaving an area. Count sites should be carefully selected in order to support one or more of the following aggregations of count data:

- aggregation across a cordon around an area to estimate total volume entering and leaving that area or
- aggregation across a screen-line to estimate total volume crossing a geographical barrier (e.g. a river or railway line) or into and out of an area from a particular direction (e.g. a suburb of a city).

The cordon might be tightly drawn and hence cover a limited area, for example an important destination such as an educational establishment or retail centre. A screen-line may also be local in nature and identify all movements onto and off a selected route from a certain direction, for example a seafront promenade or an off-road cycleway from one side of that route.

A well-constructed network of permanent automatic cycle counters (ACCs) is the essential requirement of any area-wide cycle monitoring regime which supports the aggregation of counts across cordons or screen-lines. In addition to automatic cycle counts, specific discrete and repeated manual counts are also of value. The following section discusses manual methods, including user intercept surveys. The section after that discusses the technology and applications for detecting cycle traffic using either surface or subsurface equipment. The final section discusses above-ground detection.

9.2.2   Manual classified counts

A manual count involves recording user numbers and direction of travel. In addition, a manual count can be used to determine gender and an approximate age classification (e.g. child under 16, adult, person aged 65 or older). Further than that, data could be collected on the type of cycle being ridden, the sizes of group that cyclists are in and whether luggage is being carried. Table 9.2 summarises strengths and weaknesses of manual cycle counts.

The weather and any specific conditions which might affect flows (e.g. road maintenance works) should be recorded. The Transport Research Laboratory [7] recommends counts are undertaken on at least 7 days to enable change in flow from year to year to be estimated. Sustrans, the UK Sustainable Transport charity and provider of the National Cycle Network, typically undertakes manual counts over four 12-h (7 a.m.–7 p.m.) days (Tuesday, Wednesday, Thursday and Saturday). One count period per season, performed at the same time each year, is recommended, but see Section 9.3 for further discussion on the aggregation of short-period counts to annual totals.

A user intercept survey can provide additional contextual data to a manual count. A user intercept survey data can confirm gender, age band, origin and destination, journey purpose and also journey frequency, which may be useful in factoring the count to an annual estimate. Other question topics could include general user

Table 9.2   Strengths and weaknesses of manual cycle counting

Strengths	Weaknesses
• Can be performed in locations not suitable for the installation of subsurface or above-ground automatic cycle counters (e.g. as a result of the cycleway surface) • A single surveyor can cover multiple and complex movements through a junction • A team of surveyors can provide comprehensive coverage for a small area or large junction	• Expensive because of the number of surveyor hours required • Some cycle movements could become temporarily visually screened from the surveyor, for example by passing pedestrians or motor traffic • Cycle volumes may be periodically high resulting in the potential for undercounting (e.g. at school leaving times) • The confidence interval for the count can be high as a result of human error, for example caused by distraction or boredom • The count is subject to disruption due to weather or unplanned incidents • Manual counts require repetition on a regular basis to avoid biases due to season and weather, and because of natural high variability, particularly for low cycle traffic volumes • User classifications can be systematically biased because of surveyor subjectivity

perceptions and specific questions, for example about aspects of a newly installed cycle scheme, such as the surfacing or the direction signing. If the intercept survey is undertaken on only a sample of the passing cyclists, then a complete manual count survey must also be undertaken to enable factoring.

9.2.3   Surface and subsurface equipment

Surface and subsurface equipment is classed as an ACC. Table 9.3 summarises the strengths and weaknesses of ACCs.

Surface and subsurface detection and counting systems for motor traffic are well developed and their accuracy levels well understood. Such systems typically involve detecting the passage of a large, metallic object through a magnetic field generated from coils of wire looped around a narrow rectangular slot cut into the road surface through which is passed a low-voltage electric current. The slots are back-filled with bitumen and the loop ends terminated in a cabinet or pit at the roadside that contains the data logger and battery. The battery may be changed at intervals or charged from the mains, a solar panel and/or a wind turbine.

Induction-loop-based counting technology has been adapted to count bicycles by using specific shapes of loop which account for the typically shorter and narrower cycle, and the higher level of detection sensitivity necessary because of the small mass of the cycle. The analysis software has also been adapted to allow other similar shaped objects such as prams and shopping trolleys to be differentiated and excluded from the cycle count. Different shaped loops have been developed by specific counting system manufacturers to best suit their detection algorithms with the objective of maximising the ability to detect a bicycle and determine its direction of travel.

Table 9.3   Strengths and weaknesses of automatic cycle counters

Strengths	Weaknesses
• Able to collect continuous usage data over a long period and therefore account for the influence on cycle volumes of weather and specific incidents which might affect manual counts • The long-term running costs are low relative to the quantity of data gathered • They provide data on variation by hour of day, day of week and month of year, hence allowing time series analysis of the data	• A need for capital cost outlay • Require a high level of maintenance (e.g. a battery changing schedule and regular inspections to overcome interruptions due to vandalism) • Possible gaps in the data as a result of equipment failure or vandalism • Mis-counting as a result of poor counter placement or poor counter accuracy • Lack of additional data that could be collected from manual counts, such as gender and age • Potential for systematic mis-counting if, for example, cycle trajectories miss the detection zone, or cyclists riding two abreast are counted as one cyclist

Figure 9.1   Cycle counting data logger

Figure 9.1 shows a trial site used by Transport for London to validate loop counters. One loop of the trapezoidal shape in the middle distance in the image can count a bicycle and its direction. The speed of a cycle can be determined using multiple loops. This very busy site has been provided with loops covering both sides of a wide cycleway. The data logger is located in the footway to the right-hand side. The straight-line sensors in the foreground are piezoelectric sensors which measure the deflection in the road surface caused by an object passing across it. Again, these can be tuned to screen out movements across the strip which are not cycles. Sometimes, they may be used in conjunction with an above-ground pyro-electric sensor which detects the infrared emitted by a human to assist in differentiating pedestrians from cyclists.

Other systems used to detect and count motor traffic, and which have also been applied to cycle counting, use the changes caused to a magnetic field generated by a magnetometer buried in the centre of a cycleway or road, much in the same way as inductive loop counters. Often the data will be transmitted wirelessly to a host controller. With the advent of a greater variety of cycles, such as cycles with a child-trailer, and cargo bikes, there may be a need for manufacturers to develop their loop technology further to ensure accuracy of differentiation between these cycle types.

Temporary counts can be undertaken using surface-mounted rubber tubes that generate an air pulse when a vehicle's wheels pass over the tube. The sensor records the time and sequence of the air pulses to determine the axle spacing and hence the following can be determined: direction, speed and type of vehicle (i.e. vehicle class).

Thinner gauge tubing allows the weaker air pulse generated by a cycle still to be detected. Hyde-Wright [8] and Brosnan [9] discuss the strengths and limitations of tube counters for cycle detection.

Subsurface and surface counting systems can count cycles relatively accurately and reliably. However, they do have limitations as follows:

- They can struggle to cope with multiple cycles simultaneously passing the sensors.
- The detection zone is fixed but cycles can bypass the zone if it is not located correctly, or if aspects of the cyclist's trajectory are affected temporarily or permanently by changes in the environment, for example pools of rain water forming or overhanging vegetation. Figure 9.2 illustrates an example of poor placement of an inductive cycle-count loop.
- In a mixed traffic environment, if motor traffic passes the detection zone at the same time as a cycle, the presence of the cycle can be masked by the much larger mass of the vehicle. Some degree of correction for this can be applied with some systems.
- Interference from high-voltage electrical sources, such as overhead line equipment for a railway line, can affect accuracy.
- Inductive loops and magnetic detection systems may not pick up cycles with a small proportion of metal, such as carbon-framed sports cycles.

Figure 9.2   An inductive loop too far from the edge and the cyclists' wheel paths

- Resurfacing or utility works can easily damage detectors or cabling and associated equipment.
- Installation and repairs require full closure, or significant closure, of the installation area.

9.2.4 Above-ground detection

The use of above-ground detection systems can avoid some of the limitations of surface and subsurface equipment, and they also have other advantages. The two main systems used are radar and vision-based systems. While microwave and infrared detection are used in cycle detection to activate traffic signals, they are not generally an accurate counting mechanism as they are limited by their inability to differentiate multiple cycles within the detection zone.

A radar detection system can detect an object and also measure its speed using the Doppler effect (which is an apparent shift in frequency of a wave travelling between an object and an observer as a result of the object moving either closer to or further away from the observer). A radar system will emit at a certain frequency, detect an object from the reflected wave and estimate its speed from the frequency shift of the reflected wave. Such systems can also detect the length and direction of travel of objects. Self-contained radar units can be deployed at the side of the road or cycle route at a height which provides an unobscured view of approaching cycles, and they can provide accurate data if the view remains unobscured, and groups of cyclists are not too closely packed. The system can be powered from a battery or solar power and will detect all types of cycle regardless of the material from which they are constructed.

Vision-based detection and counting systems offer a flexible and accurate method of counting cycles. At the simplest level, this involves the deployment of a small camera fixed to an item of street furniture which records images to a memory card for a specific survey period. Figure 9.3 illustrates a video survey camera installation on a piece of street furniture.

Images are downloaded and counted by enumerators in a controlled environment. The quality of the human manual count is enhanced because checks and quality control processes are able to be undertaken asynchronously from the time of the count event. The counts would usually be undertaken for the same periods as for manual classified counts (see Section 9.2.2). However, the opportunity for faster than real-time video playback means that longer periods than for equivalent manual counts can be undertaken cost-effectively if flows are low. In addition to forming the basis for a count of cyclist users, video footage can be annotated to illustrate cyclist movements of interest.

Plotting the movements on an image can provide richer insights into potential conflicts and hence the overall safety of a junction or route. The main drawbacks of video-based counting are first problems of identifying cycles during darkness or if lighting is poor, and second, that, if powered by battery, change will be required after 1–2 days. There are also challenges for the safe and secure handling of image data to ensure data protection requirements are met.

Figure 9.3   A traffic survey video camera

Table 9.4 summarises the strengths and weaknesses of video-based surveys in comparison with manual counts.

A test recording should be conducted to determine whether the camera position captures the movements required, whether there is sufficient light from street lighting

Table 9.4 Strengths and weaknesses of video-based surveys as compared with manual counts

Strengths	Weaknesses
• Possible to fast-forward through footage and therefore cover a period of time quicker than if conducting a manual count in person • Possible to rewind and pause footage to verify counts • There may already be pre-existing cameras at the scheme location (i.e. surveillance cameras) • Possible to identify the trajectories cyclists adopt	• Lighting conditions and camera quality can impair the identification, and therefore counting, of cyclists • Comparing usage at pre- and post-scheme implementation stages requires that the camera is set up in exactly the same location and at the same angle, which may not always be feasible • It may not be possible to identify demographics as easily from the video as compared with a manual count

or other lighting sources during the hours of darkness, and, conversely, whether the Sun creates glare.

More advanced vision-based systems use embedded artificial intelligence to count and classify objects within the sensor itself. These systems are trained using machine learning with a library of thousands of tagged and classified images to recognise different classes of road user, including cyclists and pedestrians, in a variety of lighting and angle of view situations. The sensors can be connected to a street lighting/traffic signal power supply or they can be solar powered to provide long-term cycle flow data. While still requiring a certain level of ambient light and a continuous power supply, embedded artificial intelligence avoids data protection issues as images are deleted almost instantly and are not transmitted beyond the sensor. Jackson *et al.* [10] describe the use of artificial intelligence vision-based cycle counting on the UK national network of canal towpaths, which are available for cycle use. Figure 9.4 shows a typical vision-based artificial intelligence sensor installation.

The same machine learning system that is built into a vision sensor can be used to post-process video footage gathered by a standard video camera. A further advantage of the artificial-intelligence-based system are the computer-generated traces of each cyclists' path through the field of view which illustrate desire lines and conflict points. Figure 9.5 shows a selection of path traces from a vision-based processing system.

Cycle movements are shown by the concentrated series of lines closest to the kerb on each side of the carriageway and also making the few lines tracing a right turn across the image from the right to the left in front of the island with the bollard.

9.3 Procedures, protocols and analysis

9.3.1 Procedures and protocols

Procedures and protocols to derive accurate and representative data are important for the collection of any data linked with the movement of people. This is

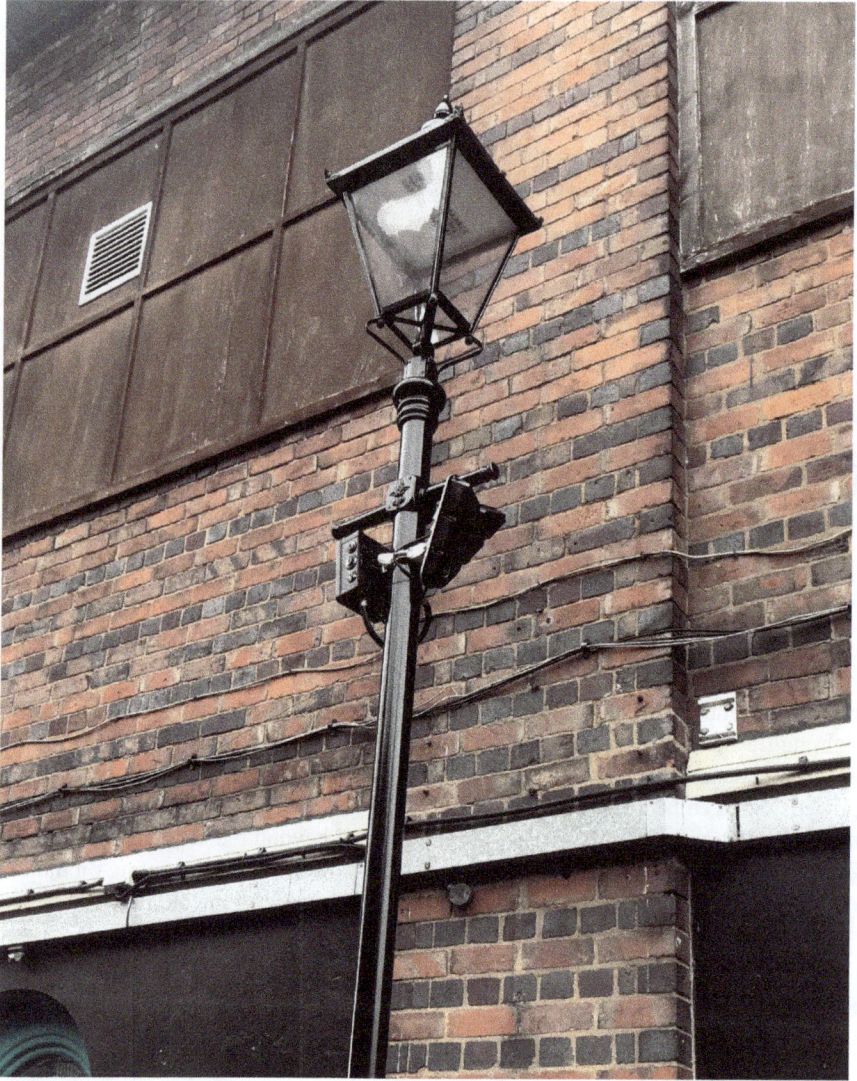

Figure 9.4   Vision-based artificial intelligence sensor installation

especially true with cycle flow data due to the particular characteristics of cycle movements described earlier, and the relative immaturity of technology for capturing data compared with motor traffic. Specific aspects to consider helping ensure the capture of a minimum quantity of appropriate data are as follows: sampling, site selection, method selection, data protection and quality control. Each of these is discussed in turn.

Figure 9.5   Path tracing from a vision-based video analysis system

The concept of cordons and screen-lines was introduced at the beginning of Section 9.2. Budget limitations may suggest that it is not possible to count at every possible cordon or screen-line crossing point. The placement of a count site at a specific location implies that the location is in some way representative for a cordon, screen-line crossing, route or area. The crossing points with the largest counts should normally be selected and this may require some preliminary manual counts to be undertaken before a permanent automatic counter is installed. If an accurate estimate of mode share is required, count sites will need to include all major vehicle crossing points as well as cycle crossing points.

As well as only a sample of sites being selected for counting, the duration of the counts may need to be limited for budget reasons. If the counts are not continuous, their timing and duration again will need to be determined based on the variability in the data and the purpose of the count. If an estimate of usage for a full year is required, counts will need to be undertaken on sufficient representative weekdays and weekend days in each season which are suitable for factoring to the period of a year. These considerations may not be so important if, for example, the data is to be used for operational appraisal: counts at peak times may suffice.

Detailed survey site assessment is critical prior to committing to the installation of equipment. Health, safety and security need to be considered in relation to the installation, maintenance and operation of survey equipment and the personnel installing and maintaining it. Consideration also needs to be given as to how health and safety issues relate to members of the public who will be passing the survey site during its operation. Pre-survey observations of cyclists' behaviours, and how these may vary as the environment changes through the survey period, should also be considered. For example, cyclists may choose a different route when it is dark or wet. Stakeholders should be consulted at every stage in the process to ensure a smooth installation and survey. These may include people owning property that fronts onto the route being surveyed, statutory undertakers (i.e. owners of electricity, gas, water and other surface and underground infrastructure within the carriageway), the police, the transport department of the local authority and other relevant departments in the local authority.

In the process of choosing a suitable survey site, consideration should be given to the choice of the most appropriate survey methods and technology. In a cycle-only environment where cyclists pass in single file, for example at a constricted point, an inductive loop counter may be effective, whereas in mixed traffic a vision-based system will provide more accurate data than other systems.

Prior to the collection of any data, a privacy impact assessment should be undertaken as a useful way of establishing what, if any, privacy issues may be relevant. Consideration needs to be given to data protection for every survey method which involves the collection of personal data. This includes user intercept surveys, and any surveys which collect and store images from video-based systems. In such circumstances, public notices relating to the capture of personal information should be displayed with the equipment informing the purpose of the data capture and a point of contact should be provided for any queries. The handling of any personal data should be the subject of a carefully managed process to ensure only legitimate access to the data for the purpose for which it is needed. The data should be destroyed after it has been used for its intended purpose.

As part of the process of installing any type of cycle-counting system, site calibration and verification should be undertaken. This will allow the accuracy of the system to be understood and any correction factors applied. All equipment requires maintenance which can include periodic battery and data storage changes, lens cleaning in the case of video and vision-based systems and clearance of overhanging vegetation. The identification of faults can also be determined during a maintenance visit or during data processing when faults can be identified through abnormal or missing data patterns. It is also possible to receive automated notifications based on live data as a result of changes in patterns compared to historic figures and changes in view or angle of camera.

9.3.2   Analysis

The data generated by subsurface and surface counters is in a variety of formats according to the equipment manufacturers' specifications. Counter equipment usually comes with software to process the data. In most cases, the process of extraction through the software aggregates the data into 15-min periods by direction as the smallest level of disaggregation. The software will then also typically report weekday and every day averages by month. Some manufacturing systems store the raw data and hence allow the analyst to aggregate in any way they may choose. Figure 9.6 shows an example of a time series plot of data from an ACC.

It can be seen by eye that the count has increased year on year, that there is significant seasonal variability, and significant variability from day to day. There are also a number of noticeable interruptions in the data series. Changes in use from year to year can be calculated by comparing the annual average daily traffic from 1 year to the next. For evaluation purposes, total annual counts may be needed, which are derived simply by factoring the average daily usage by 365.

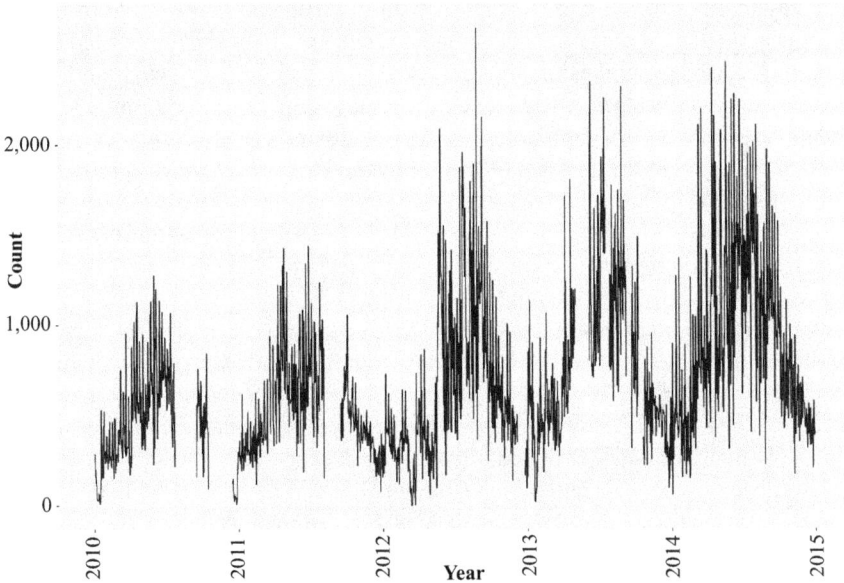

Figure 9.6   Time series plot of daily total cycle count

Cope *et al.* [11] noted that the relatively low numbers of cyclists evident in counts pose particular issues for accurate estimation of annual averages. As with all types of traffic, flow varies by hour of day, by day of week and also by month of year. In addition, there are special days, such as bank holidays, when flows may be abnormally low (e.g. on commuter routes) or abnormally high (on routes used predominantly by leisure traffic). Cycle traffic can exhibit great variability within the day; for example, on some significant commuter corridors there is more pronounced use during the peak commuting times than there is for motor traffic. Also, there is a greater variability by season than there is for motor traffic. As a separate effect to season, the weather can have a significant impact on cycle counts as well. The effect is different depending on the climate with cycling being suppressed in warm climates when there are higher temperatures than normal, and with rainfall being a significant deterrent in temperate climates. Finally, there will be different patterns of variability within the day, week and year as a function of the predominant type of use of a route (commuter, leisure or mixed).

This variability, coupled with the generally low counts, creates wide confidence intervals for estimates of the count. The effect is compounded if short-period counts are factored using expansion factors to daily or annual totals. Moreover, if the purpose of the counting is to estimate change in level of use, for example as a result of an intervention, then there is a substantially reduced probability of being able to detect statistically significant differences in either level of use, or trend in use.

Counts form a time series, and it is important to model the correlations within the data structure for counts, which may include correlations to the next day, the same day in other weeks and the same month in other years. The problem is exacerbated as a result of the count data being discrete and often small in absolute size. Gordon [12] used general estimating equations to model day of week, month, year, special day and weather effects and the underlying autocorrelation. Different models were created for commuter and leisure routes based on distinct patterns revealed through cluster analysis [13,14]. Etienne [15] used similar approaches to analyse Paris cycle hire scheme usage.

Using bicycle hire data, Fournier *et al.* [16] have developed a sinusoidal model with a single parameter to estimate monthly average daily bicycle counts and average annual daily bicycle counts. El Esawey [17], on the other hand, considers the significant problem of missing data in the time series. When missing at random, the effect on average annual daily bicycle counts was minimal even with high missing rates, but there is a greater effect on accuracy for monthly average daily bicycle counts. The difficulty is knowing whether missing data is as a result of a systematic issue, or whether the data is missing at random.

Nordback *et al.* [18], based on a study in Boulder, Colorado, estimated errors in the average annual daily bicycle counts for different counting scenarios by comparing to a continuous bicycle count. A 4-week count had an estimation error of 15% and a 1-h count an error of 54%. The suggested most cost-effective short-duration count to estimate an annual average is a period of one full week. Findings such as these, however, are likely to be relevant only for their specific locations as a result of the potential for significantly different variability in other locations. Practitioners should undertake their own local analyses using long-period automatic counts to developed local knowledge about the period appropriate for short-period cycle counts.

9.4   Innovations in cycle-counting technology

9.4.1   Harvesting digital crowdsourced data

The techniques described in Section 9.2 relate primarily to the gathering of data on cycle flows and behaviour at specific points within a network of cycle routes. A number of other techniques are available to help provide data and insight into cyclists' routing patterns and behaviour across the wider network.

Intercept interview surveys can provide detailed information about cyclists' origins and destinations, routes, motivations and perceptions. Determining the survey locations, managing the surveys at those locations in a safe manner and obtaining a representative sample by age, gender and cyclist type, such that they can be expanded using contemporaneous continuous counts, are challenging issues in the survey design. For these reasons, and also the cost of intercept surveys, alternatives that can collect much larger volumes of data are desirable.

Alternatives to intercept surveys involving automated techniques can involve either the deployment of sensors at specific locations or the harvesting of data from open data, crowdsourcing or commercial big-data sources.

Sensors can be deployed at specific points along a network which will detect the unique identifier associated with the Bluetooth or Wi-Fi chip in mobile devices such as phones and tablets carried by anybody passing the sensor. By matching an anonymised version of the identifier between survey points, a chain of points can be created which are along the route of the trip. The associated time stamps allow for travel times to be calculated. The equipment used to collect these data can be battery powered or powered from external sources and has the advantage of being able to be left in place for several weeks at a time. However, the devices are best located in cycle-only environments because of the wide and overlapping confidence intervals associated with the speed distributions of the different modes of the carrier. These overlapping speed distributions give high error rates in associating a portable device carrier with a particular mode. The sample size will be limited by the level of detail that a detection zone can be set to, the number and spread of sites surveyed and the number of detectable devices in use. Currently, typically fewer than 10% of cyclists within a flow will be carrying a detectable device, but some people may carry multiple devices.

At a larger scale, even up to the level of a whole conurbation, it might be possible in the future to infer trips by cycle from mobile phone network data. These data sets are derived from the on-call and off-call communications made between a mobile phone and the cell masts of the mobile phone network operator. By analysing the movement of phones between masts and applying algorithms to infer mode of travel and trip purpose, it is possible to build a picture of origin–destination movements using millions of items of trip information. The accuracy of the data is limited as a result of the mode being inferred from the speed profile. Data protection rules mean that data is aggregated to avoid personal identification and may be for periods longer than, for example, a single peak hour, or aggregated over a number of days. It is sometimes not possible to know the ultimate origin and the ultimate destination as trips may be truncated by the data provider to prevent knowledge about the true origin and destination locations of trips. A further issue concerns the potential variable penetration of the network providers in the market, and the issue of a person carrying multiple devices (e.g. a personal and a work phone). Mobile phone network data can, however, be validated and expanded by comparison with other data sets such as household survey data, census data or transport user intercept data.

Trip patterns can also be investigated using data generated from the GPS and accelerometer functions integrated into mobile phones and other equipment such as connected watches and cycle lights [19]. The use of this functionality by cyclists through mobile platform apps, such as fitness trackers and community crowd-sourced data apps, allows data sets to be downloaded and analysed to provide information on trip patterns and behaviour. Consideration needs to be given to the representativeness of the sample due to the motivations behind using the apps and the limitations placed on it by the need to protect personal information.

9.4.2 *Issues and a future trajectory*

Transport system operators and people on the move are generally becoming more connected to the internet. This will increasingly provide opportunities for surveying transport use within the transport system. Data from cycle hire schemes, for example, can already be used to understand patterns of use. It is typically the case that countries with high volumes of cycle use also have a more diversified fleet of cycles, including, for example, cycles used to carry children and cargo cycles. The journeys may hence also not only increase in number but diversify in terms of their trip purpose. Cycle journeys may also become longer as a result of greater market penetration of electrically assisted cycles.

The way we travel in the future using Mobility as a Service technology and intelligent connected highway infrastructure will generate masses of movement data. Data sets specific to cycling may result from connected infrastructure communicating with cycles and cyclists to provide priority and safe passage through a network. Data sets of this type are already starting to be generated from bike share schemes [15].

Acknowledgements

We would like to acknowledge the valuable review comments provided by George Macklon, Fiona Crawford and Paul Frões.

References

[1]   Transport for London (2017) Travel in London Report 10. Transport for London. Available at http://content.tfl.gov.uk/travel-in-london-report-10.pdf. Accessed on 27/07/18.

[2]   Marqués, R and Hernández-Herrador, V (2017) "On the effect of networks of cycle-tracks on the risk of cycling. The case of Seville", Accident Analysis and Prevention, vol. 102, pp. 181–190.

[3]   Sloman, L, Goodman, A, Taylor, I, *et al.* (2017) Cycle City Ambition Programme: Baseline and Interim Report, for UK Department for Transport. Available at https://assets.publishing.service.gov.uk/government/uploads/system/uploads/attachment_data/file/738307/170912-cycle-city-ambition-stage-2-baseline-report-final.pdf. Accessed on 30/10/18

[4]   Cope, A, Carter, S, Buglass, E, and Macklon, G (2015) Cycle City Ambition Monitoring Plan, for UK Department for Transport. Available at https://assets.publishing.service.gov.uk/government/uploads/system/uploads/attachment_data/file/486384/cycle_city_ambition_grant-report.pdf. Accessed on 30/10/18.

[5]   Parkin, J (2018) Designing for Cycle Traffic: International Principles and Practice. ICE Publishing, London.

[6]   Transport for Greater Manchester (2016) Transport Statistics Report. Available at http://www.gmtu.gov.uk/reports/transport2016.htm. Accessed on 30/10/18.

[7]   Davies, DG, Emmerson, P, and Pedler, A (1999) Guidance on monitoring local cycle use. Transport Research Laboratory Report 395. Transport

Research Laboratory, Crowthorne. Available at https://trl.co.uk/sites/default/files/TRL395.pdf. Accessed on 29/10/18.

[8] Hyde-Wright, A, Graham, B, and Nordback, K (2014) "Counting bicyclists with pneumatic tube counters on shared roadways, Institute of Transportation Engineers", ITE Journal, vol. 84, p. 2; Social Science Premium Collection.

[9] Brosnan, M, Petesch, M, Pieper, J, Schumacher, S, and Lindsey, G (2015) "Validation of bicycle counts from pneumatic tube counters in mixed traffic flows", Transportation Research Record: Journal of the Transportation Research Board, vol. 2527, pp. 80–89.

[10] Jackson, P, Clarke, H, Reeves, G, and Rutter, R (2018) Waterways and Wellbeing: Building the Evidence Base. Transport Practitioners Meeting, Oxford, 5–6 July 2018.

[11] Cope, AM, Abbess, CR, and Parkin, J (2007) Improving the empirical basis for cycle planning. Paper in Mathematics on Transport, Selected Proceedings of the 4th Institute of Mathematics and its Applications International Conference on Mathematics in Transport. Elsevier.

[12] Gordon, G (2013) Investigating methodologies for analysing single point count data to estimate volumes of bicycle traffic. PhD Thesis, University of Bolton.

[13] Gordon, G and Parkin, J (2012) "Patterns of use by season, day of week and time of day that lead to identifying distinct cycle route typologies", Cycling Research International, vol. 2, pp. 104–18.

[14] Gordon, G and Parkin, J (2012) Developing methodological approaches to analysing single point bicycle counts. The fourth annual Australian Cycling Conference, Adelaide, 16–17 January 2012.

[15] Etienne, C and Latifa, O (2014) "Model-based count series clustering for bike sharing system usage mining: a case study with the Vélib' system of Paris", ACM Transactions on Intelligent Systems and Technology (TIST), vol. 5, no. 3, pp. 1–21.

[16] Fournier, N, Christofa, E, and Knodler, MA (2017) "A sinusoidal model for seasonal bicycle demand estimation", Transportation Research Part D, vol. 50, pp. 154–169.

[17] El Esawey, M (2018) "Impact of data gaps on the accuracy of annual and monthly average daily bicycle volume calculation at permanent count stations", Computers, Environment and Urban Systems, vol. 70, pp. 125–137.

[18] Nordback, K, Marshall, WE, Janson, BN, and Stolz, E (2013) "Estimating annual average daily bicyclists: error and accuracy", Transportation Research Record: Journal of the Transportation Research Board, vol. 2339, no. 1, pp. 90–97

[19] SeeSense (2018) Safer cycling for sustainable cities. Available at https://seesense.cc/pages/smart-cities. Accessed on 20/10/18.

Chapter 10

Crowd density estimation from a surveillance camera

Viet-Quoc Pham[1]

This chapter presents an approach for crowd density estimation in public scenes from a surveillance camera. We formulate the problem of estimating density in a structured learning framework applied to random decision forests. Our approach learns the mapping between image patch features and relative locations of all the objects inside each patch, which contribute for generating the patch density map through Gaussian kernel density estimation. We build the forest in a coarse-to-fine manner with two split node layers and further propose a crowdedness prior and an effective forest reduction method to improve the estimation accuracy and speed. Moreover, we introduce a semiautomatic training method to learn the estimator for a specific scene. We achieved state-of-the-art results on the public Mall and UCSD datasets and also proposed two potential applications in traffic counts and scene understanding with promising results.

10.1 Introduction

Counting objects in images is important in many real-world applications, including traffic control, industrial inspection, surveillance, and medical image analysis. Counting in crowded scene is nontrivial owing to severe inter-object occlusion, scene perspective distortion, and complex backgrounds. Besides counting, estimating crowd density is necessary for understanding crowd behavior, especially in large public spaces such as train stations and streets. This chapter addresses the problem of counting in public scenes based on crowd density estimation, as shown in Figure 10.1.

Existing counting methods can be classified into three categories: counting by detection [1,2], counting by clustering [3,4], and counting by regression [5–10]. In the first two categories, counting is based on individual detection and motion segmentation that are sensitive to heavy object occlusion and cluttered background. Particularly, for such crowded scenes that are only a part of object instance can be observed, detection and segmentation of individuals become impracticable. In counting by regression, counting techniques learn a mapping between low-level

[1]Corporate Research and Development Center, Toshiba Corporation, Tokyo, Japan

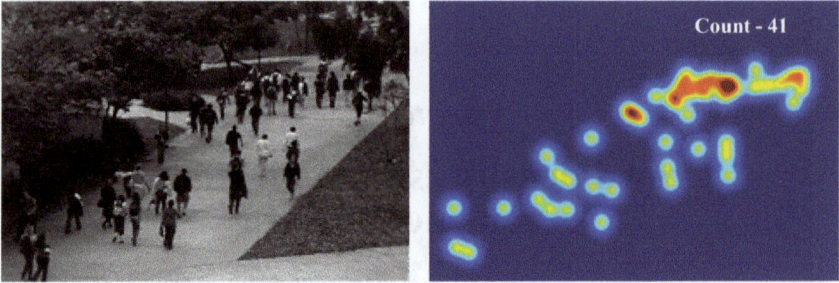

Figure 10.1   Our objective is to estimate the object density map and the number of objects (right) from an input image (left)

features and people count, and therefore they can avoid explicit object detection and segmentation in the crowded scenes.

Counting by regression methods differs depending on the target of regression: the object count, or the object density. Chen *et al.* [5,6] learn the regression mapping between the low-level imagery feature and the number of objects. They achieved state-of-the-art results for counting people by introducing the concept of a cumulative attribute to the regression problems, which improved the counting accuracy by considering the cumulative, dependent nature of regression labels [5]. In another approach, Lempitsky and Zisserman [7] estimate an image density who's integral over any image region gives the count of objects within that region. The object density is more informative than the object count, since it can give a *rough estimate on the object locations*, as shown in Figure 10.1. Therefore, crowd density estimation is more useful for understanding the crowd behavior. To estimate the crowd density, Lempitsky and Zisserman learn a linear transformation of the feature representation $f(x)$ that approximates the density function at each pixel $D(x) = \omega^T f(x)$ [7]. The problem here is that it is difficult to design a feature satisfying the linear approximation hypothesis.

This chapter presents a patch-based approach for crowd density estimation in public scenes. Different from the work of Lempitsky and Zisserman [7] assuming a linear transformation, we aim to learn the nonlinear mapping between patch features and relative locations of all objects inside each image patch using a random forest framework. Object locations estimated in each patch will contribute for generating the patch density map through Gaussian kernels. The patch feature used in our method is much simpler and scene independent. We name our model *COUNT forest* (CO-voting Uncertain Number of Targets using random forest), as it uses random forest regression from multiple image patches to vote for the densities of multiple target objects. We build the forest in a coarse-to-fine manner with two split node layers and train a prediction label at each leaf node. One of the advantages of the COUNT forest model is that it requires much less memory to build and reserve the forest, compared with regression forest models with densely structured labels proposed in [8,11]. This is a prerequisite for an embedded computer vision system.

As there is a large variation in appearance and shape between crowded image patches and non-crowded ones, we propose a *crowdedness prior*, which is a global property of the image, to train two different forests corresponding to this prior. We develop a robust density estimator by adaptively switching between the two forests. Moreover, we also propose a nontrivial approach of effective forest reduction by using permutation of decision trees. This improvement speeds up the estimation so that it can run in real time. Another contribution is the introduction of a semiautomatic training method to learn the estimator for a specific scene. We synthesize training samples randomly from a large set of segmented human regions and the target scene background. The synthesized training samples not only facilitate labor-saving but also adapt our estimator to the target scene.

Finally, we verified the performance on various crowded scenes. We achieved state-of-the-art results on the public Mall [6] and UCSD datasets [9] and confirmed high performance for multiple-class counting and scene adaptation with good results for the Train Station dataset [12] and a newly captured dataset. We also proposed two potential applications in traffic counts and scene understanding with promising results. The conference version of this chapter can be referred to from [13].

We discuss related works in Section 10.2 and explain the proposed COUNT forest model in Section 10.3. Improvements of the method, including the crowdedness prior, forest permutation, and semiautomatic training, are explained in Section 10.4. We present experimental results in Section 10.5.

10.2 Related works

Counting by regression: Chen *et al.* [5,6] learn the regression mapping between the low-level imagery feature and the object count. In [6], they propose a single multi-output model based on ridge regression that takes interdependent local features from local regions as input, and people count from individual regions as multi-dimensional structured output. They further introduce the concept of the cumulative attribute for regression to address the problems of feature inconsistency and sparse data [5]. In this work, they convert the people count into a binary cumulative attribute vector and use it as the intermediate representation of the image to learn the regression models. Loy *et al.* [10] develop a framework for active and semi-supervised learning of a regression model with transfer learning capability, in order to reduce laborious data annotation for model training. Lempitsky and Zisserman [7] model the crowd density at each pixel, casting the problem as that of estimating an image density whose integral over any image region gives the count of objects within that region. This model is effective, but the linear model requires a scene-dependent and relatively complex set of features (BoW-SIFT).

Random forest: Part-based object detection methods learn the appearance of object parts and their spatial layout to model an object. Okada [14] and Gall and Lempitsky [15] proposed a random forest framework [16,17] for learning a direct mapping between the appearance of an image patch and the object location. The combination of the tree structure and simple binary tests makes training and matching against the codebook very fast, whereas clustering-based learning of explicit codebooks is considerably more expensive in terms of both memory and time. The idea of probabilistic voting was also exploited in object tracking [18], action recognition [19], facial feature detection [20], and human pose estima-tion [21]. The main difference between the COUNT forest and other voting-based methods is the output of regression. In our method, each patch votes for locations of nearby objects, whose quantity and structure are not fixed due to the large variation of crowds.

10.3   COUNT forest

Our problem is that of estimating a density map of target objects as shown in Figure 10.1, given a set of training images with annotations. The annotation is a set of dots located at centers of object regions, as shown in Figure 10.2. Dotting is a natural way for humans to count objects, at least when the number of objects is large, and it is less laborious than other annotation methods such as bounding boxes. For a training image I, we define the ground truth density function for each pixel $p \in I$ to be a kernel density estimated based on the dots provided:

$$F^0(p) = \sum_{\mu \in A} \mathcal{N}(p; \mu, \sigma^2 \mathbf{1}_{2 \times 2}),\tag{10.1}$$

where $\mathbf{A}$ is a set of user annotation dots and $\mathcal{N}(p; \mu, \sigma^2 \mathbf{1}_{2 \times 2})$ denotes a 2D Gaussian kernel centered on each dot $\mu \in \mathbf{A}$ with a small variance ($\sigma = 2.5$ pixels).

We aim to learn a nonlinear mapping $\mathcal{F}$ between patch features v and locations l of all objects inside each image patch relative to the patch center:

$$\mathcal{F} : v \in \mathbf{V} \to l \in \mathbf{L},\tag{10.2}$$

Figure 10.2   User annotation dots at people heads (target objects)

where $\mathbf{V}$ is the feature space and $\mathbf{L}$ is the label space. The patch feature v is a vector concatenating responses of feature channels at every pixel inside the image patch. We augment each image patch with multiple additional filter channels, resulting in a feature vector $v \in \mathbb{R}^{h \times w \times C}$ where C is the number of channels and $h \times w$ is the patch size. A label l denotes a set of displacement vectors from the patch center O to all object locations P_j within its neighbor region of radius R:

$$l = \{\vec{OP_j}|P_j \in A \wedge |\vec{OP_j}| < R\}. \tag{10.3}$$

Some samples of l are shown in Figure 10.3. We propose a novel random forest framework, which we call *COUNT forest*, to learn the mapping $\mathscr{F}$ and use this forest regression from multiple image patches to vote for the densities of multiple target objects. The biggest advantages of our COUNT forest model are that it requires small memory space to build and reserve the forest, as it contains only vector labels at its leaf nodes. Compared with regression forest models with densely structured labels proposed in [8,11], our model is more suitable for embedded computer vision systems.

10.3.1   Building COUNT forest

Our COUNT forest is an ensemble of randomized trees that classify an image patch by recursively branching left or right down the tree in a coarse-to-fine manner until a leaf node is reached, as shown in Figure 10.3.

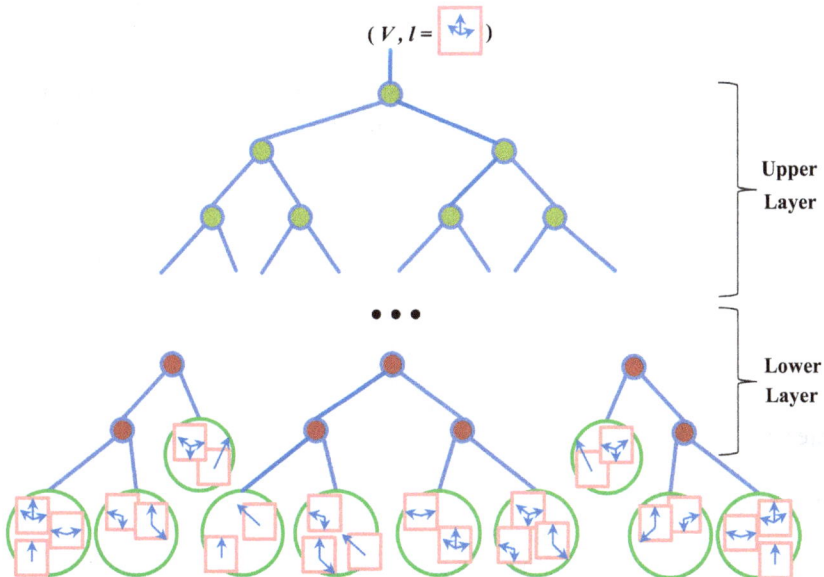

Figure 10.3   Training COUNT forest in a coarse-to-fine manner using two split layers. For space limit, only one tree is shown

Training of each tree proceeds as follows: given a training set $\{(v_i, l_i)\}$, we assign all the training samples to the root node and recursively repeat splitting of the node until the number of training samples in a node becomes small or the height of the tree becomes large. For each node, we choose J splits randomly, each of which is a pair of a test function and a threshold. All data $S_j = \{(v_i, l_i)\}$ arriving at node j are split into a left subset $S_{L(j)}$ and a right subset $S_{R(j)}$ based on the thresholding result of the test function:

$$
\begin{aligned}
S_{L(j)} &= \{(v_i, l_i) \in S_j | f_j(v_i) < t_j\} \\
S_{R(j)} &= S_j \backslash S_{L(j)},
\end{aligned}
\tag{10.4}
$$

where $f_j(v)$ and t_j are the test functions of the feature vector v and the threshold at a node j, respectively. In our implementation, $f_j(v)$ is a randomly selected element of the feature vector v.

Because our regression target is the vector label that differs by both the number of displacement vectors and their spatial distribution, we found that a single split criterion used in previous works is not enough to classify such labels. Our solution is to apply a coarse-to-fine approach where we separate the split nodes into *two split layers* with a different split criterion. In the upper split layer, we attempt to split the labels based on their spatial distributions. Here, we find the best split (f_j^*, t_j^*) by minimizing the total variance:

$$
(f_j^*, t_j^*) = \underset{(f_j^n, t_j^n)}{\arg\min} \sum_{o \in S_{L(j)}} ||\mathbf{H}^o - \bar{\mathbf{H}}^L||_F^2 + \sum_{o \in S_{R(j)}} ||\mathbf{H}^o - \bar{\mathbf{H}}^R||_F^2,
\tag{10.5}
$$

where $\mathbf{H}^o$ is a 2D histogram representing the spatial distribution of object locations indicated by the label, as shown in Figure 10.4. Each histogram bin of $\mathbf{H}$ is computed from contributions of nearby object locations via Gaussian kernels. $\bar{\mathbf{H}}^L$ and $\bar{\mathbf{H}}^R$ are the average histograms of $\mathbf{H}$'s belonging to the left subset $S_{L(j)}$ and the right subset $S_{R(j)}$, respectively, after splitting. $|| \cdot ||_F$ is the Frobenius norm. When the maximum depth or the minimum node size of this split layer is reached, patches are then passed to the lower split layer.

In the lower split layer, we attempt to make finer splits based on the label size, which is defined as the number of displacement vectors, i.e., the number of object locations, indicated by each label. We use the class uncertainty measure as the split function:

$$
(f_j^*, t_j^*) = \underset{(f_j^n, t_j^n)}{\arg\min} \frac{|S_{L(j)}|}{|S_j|} \mathcal{H}(S_{L(j)}) + \frac{|S_{R(j)}|}{|S_j|} \mathcal{H}(S_{R(j)}),
\tag{10.6}
$$

where $\mathcal{H}(S)$ is defined as the Shannon entropy of the distribution of label sizes in S. As a result, the uncertainty of class distribution of all the labels associated with a node decreases with increasing tree depth, and each leaf should contain a set of labels with almost the same label size.

We must notice that in the test phase, we do not need to distinguish the two split layers because we use the same testing criterion for every node.

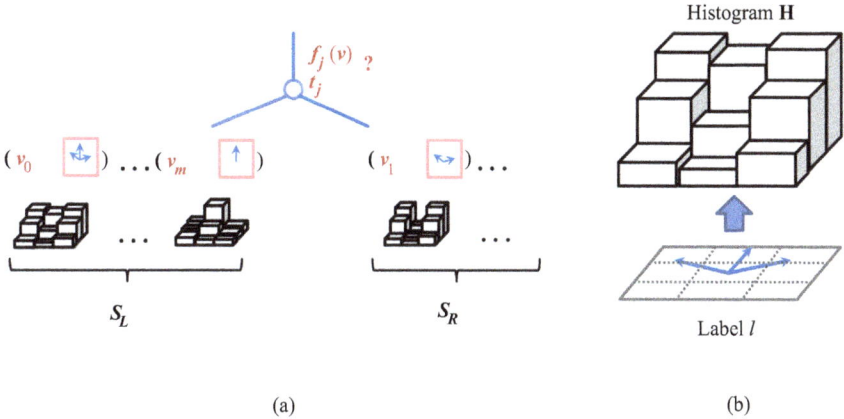

Figure 10.4 *Splitting criterion of the upper split layer (a) based on spatial distribution histograms (b)*

10.3.2 *Prediction model*

In this section, we explain how to predict the output label of the abovementioned randomized tree using the labels stored in the leaves during training time. An illustration of this prediction model is shown in Figure 10.5.

After the training process stated in Section 10.3.1, training samples are stored in leaf nodes of the forest. We divide all the labels stored in each leaf node into clusters with the same label size (see Figure 10.5). Among these clusters of labels, we select the cluster G having the largest number of members. As stated in the previous section, labels in each leaf node have almost the same size; therefore, G contains most labels in the leaf and the remaining labels can be considered as noises. We assume that the labels in G have the same label size K. Note that K varies with the labels stored in each leaf. In Figure 10.5, we show the case when $K = 2$. In the next step, we project the object locations indicated by all the labels in G on a plane whose center corresponds with each patch center of the labels and apply K-means to obtain K clusters of such locations. Finally, we create a new label consisting K displacement vectors from the plane center to the K centroids of these clusters. We call it the prediction label and substitute it for all labels in the same leaf node for making the final trained tree. The built forest consists of only vector labels in its leaf node, and therefore *its memory size is much smaller* than regression forest models with densely structured labels proposed in [8,11].

10.3.3 *Density estimation by COUNT forest*

In Figure 10.6, we explain the procedure of estimating object densities using the trained COUNT forest. The procedure starts by extracting all the patches from the input image, and a feature vector v_i from each patch i is computed. We then classify this vector v_i by recursively branching left or right down each tree T_j in the learned forest until a leaf node is reached and obtain a prediction label l_{ji} stored in this leaf node. In the standard voting procedure [15], each label votes a delta peak for each

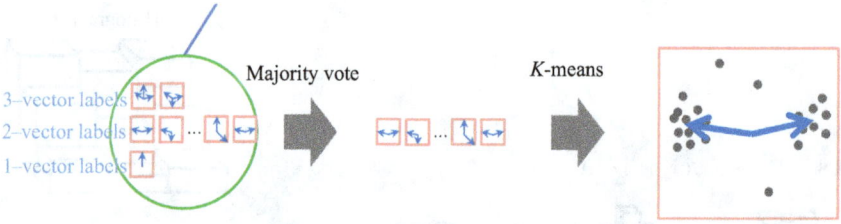

Figure 10.5 *Prediction model by majority voting and K-means. In this sample,*
K = 2

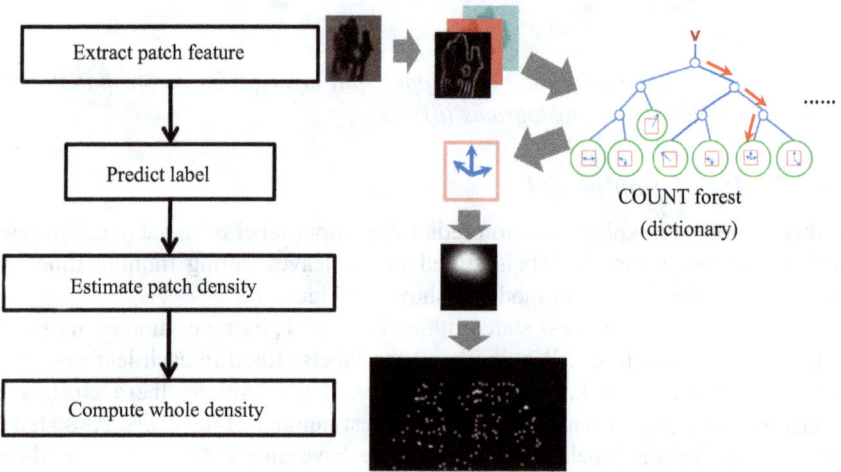

Figure 10.6 *Density estimation procedure*

object location, and the vote counts accumulated in each pixel are then Gaussian filtered to obtain a voting result map. Because each false predicted object location will increase the error of object counts by 1, we cannot use this voting map as our desired density map. Applying non-maxima suppression and counting from the object detection results do not work either because objects are so closed to each other or often partly occluded in the crowded scenes (see Figure 10.1).

Borrowing the idea of neighbor smoothing in [11], we suppress the error in estimating a density for each pixel by collecting predictions from neighboring pixels. We calculate a patch density map from a predicted label and average it across trained trees and across neighbor patches. For details, each object location indicated by a label l_{ji} contributes to the patch density map by a Gaussian kernel, following the definition of the density function in (10.1):

$$D_{ji}(x) = \sum_{\mu \in l_{ji}} \mathcal{N}(x; \mu, \sigma^2 \mathbf{1}_{2 \times 2}), \tag{10.7}$$

where x is an arbitrary pixel inside patch i, and μ is an object location indicated by label l_{ji}. Note that this patch density map has the same size as the input image patch. For fast computation, we precompute the Gaussian kernels $\mathcal{N}(x; \mu, \sigma)$ and store them in a lookup table indexed by $(x - \mu)$. The final density map of patch i is computed by averaging D_{ji} over the set of trees $\mathcal{T}$ of the trained forest:

$$D_i(x) = \frac{1}{|\mathcal{T}|} \sum_{T_j \in \mathcal{T}} D_{ji}(x). \tag{10.8}$$

The density map for the whole image is computed by averaging all the predicted overlapping density maps $D_i(x)$:

$$D(x) = \frac{1}{|\mathcal{D}(x)|} \sum_{D_i \in \mathcal{D}(x)} D_i(x), \tag{10.9}$$

where $\mathcal{D}(x)$ is the set of density maps that contain pixel x in their scope. The average computations in (10.7)–(10.9) smooth the density predicted at each pixel by incorporating neighbor pixel information, and therefore the estimation error at a single pixel can be suppressed.

10.4 Robust density estimation

In this section, we use the COUNT forest described in the previous section to develop a robust density estimator with three improvements: increasing accuracy with a crowdedness prior, speeding up estimation by an effective forest reduction method, and decreasing annotation work by semiautomatic training.

10.4.1 Crowdedness prior

In Figure 10.7, we show two examples of the crowded and non-crowded scenes. As seen in the figure, there is a large variation in appearance and shape between crowded image patches and non-crowded ones. Therefore, learning a forest from the entire training set, including both the crowded and non-crowded scenes, should be harder than learning different forests for each scene. We use this crowdedness prior, which is a global property of the image, for developing a robust density estimator.

We aim to estimate a density D_i^n of an image patch i in the current frame, given the density D_i^{n-1} estimated from the previous frame. By introducing the crowdedness prior c, the probability $p(D_i^n | D_i^{n-1})$ can be estimated as

$$P(D_i^n | D_i^{n-1}) = \sum_{j=1,2} P(D_i^n | c_j, D_i^{n-1}) P(c_j | D_i^{n-1}), \tag{10.10}$$

where c_1 and c_2 denote the crowded and non-crowded properties, respectively. The probability $p(D_i^n | c_j, D_i^{n-1})$ is independent of D_i^{n-1} and can be learned by training a COUNT forest on different training subsets. We collect crowded and non-crowded patches for training a different forest for c_1 and c_2. A patch i is called crowded if the number $\mathcal{N}_i$ of objects in its neighborhood is inside the range $[1/3\mathfrak{N}, \mathfrak{N}]$, and non-crowded if $\mathcal{N}_i \in [0, 2/3\mathfrak{N}]$, where $\mathfrak{N} = \max \mathcal{N}_i$. The prior probability $p(c_j | D_i^{n-1})$

Figure 10.7   Crowded image patches (left) and non-crowded ones (right) differ largely in appearance and shape

is defined as

$$P(c_1|D_i^{n-1}) = [\mathcal{N}_i^{n-1} > \mathfrak{N}/2], \tag{10.11}$$

$$P(c_2|D_i^{n-1}) = 1 - P(c_1|D_i^{n-1}). \tag{10.12}$$

The motivation for these simple definitions is that we need to process *only one forest* to estimate a density map for a patch. As a result, we can improve accuracy by using this crowdedness prior without increasing the calculation cost.

10.4.2   Forest permutation

As other regression forest-based methods, the speed of our density estimation method depends on the number of trees to be loaded and the sampling stride, i.e., the distance between sampled image patches. A trivial approach of fixedly reducing the forest could affect the accuracy heavily due to the weak ensemble of few decision trees.

We proposed a nontrivial approach of reducing the forest using permutation of decision trees. We divide the original forest into several sub-forests (typically 4) and circularly shift these sub-forests whenever moving to the next patch (e.g., $1234 \rightarrow 2341 \rightarrow \cdots$). We then modify the formulation (10.8) as follows:

$$D_i(x) = \frac{1}{|\mathcal{T}_{i,1}|} \sum_{T_j \in \mathcal{T}_{i,1}} D_{ji}(x), \tag{10.13}$$

where $\mathcal{T}_{i,1}$ is the first sub-forest after the forest permutation at patch i. Our idea is based on the fact that a density at a pixel is calculated by averaging predicted patches across trained trees and across neighbor locations. As shown in Figure 10.8, although each density patch is predicted using only one sub-forest, a density at each pixel is eventually computed from the whole forest, which appears partly in the surrounding patches.

10.4.3   Semiautomatic training

In this section, we address another common problem of counting-by-regression methods: manual annotation. As in other methods [5–9], we also require a different set of annotated images to train the estimator for a specific scene. Annotating dozens of images is laborious work, even in the case of a simple task such as marking the head of each person as shown in Figure 10.2.

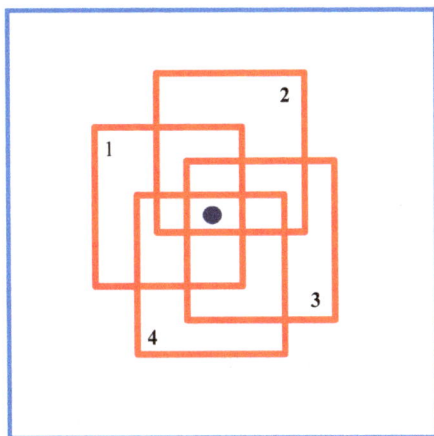

Figure 10.8   The density map for each red patch is predicted using a different sub-forest; therefore, the density at the center dot is computed from all four sub-forests

Figure 10.9   Segmentation masks obtained by GrabCut [23]

To deal with this problem, we introduce a semiautomatic training method for learning the estimator for a specific scene, making our technique more practical. Instead of selecting and annotating real images, we synthesize training images from a large set of segmented human regions and the target scene background. The synthesized training samples not only facilitate labor-saving but also adapt our estimator to the new scene. Moreover, since we deal with image patches in the crowded scenes with small object sizes, we do not need high qualities of synthetic data. As shown in our experiments in Section 10.5, our COUNT forest model is sufficiently robust to take simple synthesized data as the training samples to obtain high performance.

In our implementation, we perform image segmentation on the PETS2009-S2 dataset [22], including 795 images with bounding-box annotations, by using the GrabCut method [23]. We obtain more than 2,000 segmentation masks, as shown in Figure 10.9. The synthesizing process starts by pasting the random segmentation masks at random positions in the target scene background. We then fix the mask size according to the perspective scale of the scene and synthesize shadows for making the final augmented images.

10.5   Experiments

The parameters of our COUNT forest are set as follows: the number of trees is $|\mathcal{T}| = 32$, the tree depth is $h_{\max} = 11$, the maximum depth for the upper split layer is $h^1_{\max} = 8$, and the minimum split size is $n_{\min} = 20$. The patch size is fixed to 13×13 pixels. In our training process, we extracted 1,000 patches from each training image to build the set of training samples. We used the following feature channels for computing the feature vector v: the raw image, the background subtraction result, the temporal derivative, the Gaussian gradient magnitude, the Laplacian of Gaussian, and the eigenvalues of the structure tensor at different scales 0.8, 1.6, 3.2 (here we used the covariation matrix of derivatives over the pixel neighborhood). To account for the perspective distortion, we multiplied all feature values with the square of the provided camera perspective map.

We use the same metrics as conventional works [5,6,10] for evaluating counting performance: mean absolute error $mae = E(|\kappa_j - \widehat{\kappa}_j|)$, mean squared error $mse = E((\kappa_j - \widehat{\kappa}_j)^2)$, and mean deviation error $mde = E(|\kappa_j - \widehat{\kappa}_j|/\kappa_j)$, where κ_j and $\widehat{\kappa}_j$ are the true and the estimated numbers of objects in frame j, respectively. $\widehat{\kappa}_j$ is computed as the sum of estimated densities over the whole image.

10.5.1   Counting performance

We evaluate the performance of counting people on the two public datasets: UCSD [9] and Mall [6]. Both the datasets are provided with dotted ground truth [24,25]. The details of both the dataset are shown in Table 10.1. Sample images from the two datasets and the corresponding density maps estimated by our method are shown in Figure 10.10.

In the first experiment, we used the same experimental setting as [5,6,10]. For the UCSD dataset, we employed 800 frames (600–1,400) for training and the rest (1,200 frames) for testing. For the Mall dataset, the first 800 frames were used for training and the remaining 1,200 frames for testing.

We perform comparison against conventional methods: recent pedestrian detector (Detector [26]),[*] least square support vector regression (LSSVR [27]), kernel ridge regression (KRR [28]), random forest regression (RFR [29]), Gaussian process regression (GPR [9]), ridge regression (RR [6]), cumulative attribute ridge regression (CA-RR [5]), semi-supervised regression (SSR [10]), and maximum excess over subarrays (MESA [7]). The accuracy comparison results of counting people for the two datasets are shown in Table 10.2. We achieved the state-of-the-art performance on all the three metrics and improved the mean absolute error (*mae*) by 27% relative to the best previous result on the Mall dataset [5].

We also perform a comparison on the UCSD dataset with other four different training/testing splits of the data ("max," "down," "up," "min") as used in [7] and

---

[*]The pedestrian detector did not work for the UCSD dataset due to small sizes of people.

Table 10.1   *Dataset details: from left to right: dataset name, number of frames, resolution, frame per second, minimum and maximum number of people in the ROI, total number of people instances*

Data	Frame	Resolution	FPS	Count	Total
UCSD	2,000	238×158	10	11–46	49,885
Mall	2,000	640×480	2	13–53	62,325

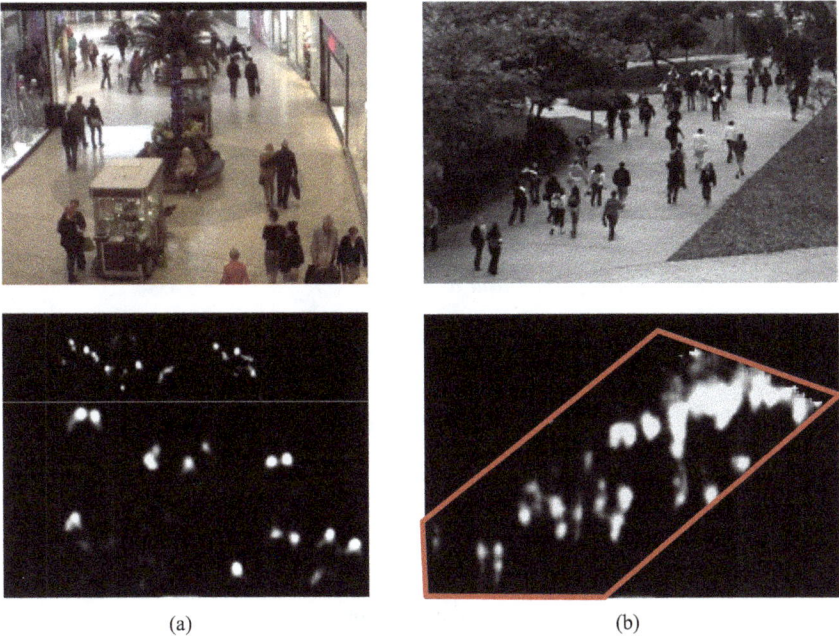

(a)   (b)

Figure 10.10   *Datasets used in our experiments: (a) Mall [6], (b) UCSD [9]. Upper row: input image, lower row: density map. The red border line indicates the ROI for density estimation*

show the result in Table 10.3. We outperform [8] for all split settings and are comparable with Arteta *et al.* [30].

To evaluate the estimated displacements of pedestrians, we propose a recall function for each image, which is the fraction of ground-truth person instances that the sum of estimated density inside its bounding box is larger than an overlap threshold. In Figure 10.11, we show the average recalls (AR) over all frames with corresponding overlap thresholds. AR at threshold 50% is 96%, meaning that 96% of persons are detected close to their correct locations.

Table 10.2   Comparison on the Mall [6] and the UCSD datasets [9]

Method	Mall [6]			UCSD [9]		
	mae	*mse*	*mde*	*mae*	*mse*	*mde*
Detector [26]	20.55	439.1	0.641	–	–	–
LSSVR [27]	3.51	18.2	0.108	2.20	7.3	0.107
KRR [28]	3.51	18.1	0.108	2.16	7.5	0.107
RFR [29]	3.91	21.5	0.121	2.42	8.5	0.116
GPR [9]	3.72	20.1	0.115	2.24	8.0	0.112
RR [6]	3.59	19.0	0.110	2.25	7.8	0.110
CA-RR [5]	3.43	17.7	0.105	2.07	6.9	0.102
SSR [10]	–	17.8	–	–	7.1	–
Ours	**2.50**	**10.0**	**0.080**	**1.61**	**4.4**	**0.075**

We achieved the best results on all three metrics.

*Table 10.3   Mean absolute errors in the UCSD dataset with four different training/
testing splits of the data as used in [7]*

Method	"max"	"down"	"up"	"min"
MESA [7]	1.70	1.28	1.59	2.02
RF [8]	1.70	2.16	1.61	2.20
Arteta *et al.* [30]	1.24	1.31	1.69	1.49
Ours	1.43	1.30	1.59	1.62

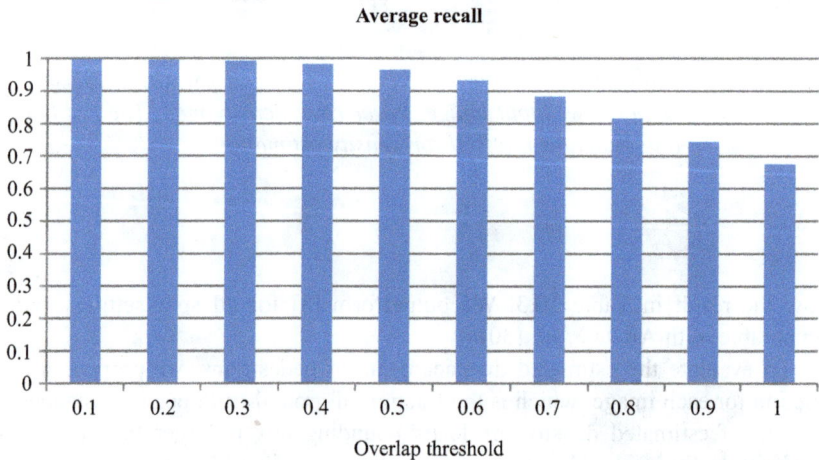

*Figure 10.11   Average recall to measure the effectiveness of person displacement
estimation in the UCSD dataset*

10.5.2   Robustness

In Table 10.4, we present detailed evaluations of the COUNT forest and the proposed improvements in Section 10.4. The methods are compared in their accuracy, speed and memory cost. All experiments were performed using C++ on a PC with two Intel Xeon CPUs (2.80 GHz). We implemented the regression forest with densely structured labels proposed in [11] as our strong baseline.[†] Our COUNT forest gives a better accuracy and the memory cost for loading the forest is 30 times smaller. When applying the crowdedness prior, we further increase the accuracy at the cost of doubling the dictionary size, as we use two forests for estimation. In Figure 10.12, we show the ground truth and estimated pedestrian counts by the two approaches. By adaptively switching between the two forests, we obtained better estimation in extremely crowded and non-crowded scenes.

We then applied the forest permutation approach in Section 10.4.2, where the sampling stride was set to 3 and the number of sub-forests was 4. There was a small loss in accuracy, but we could achieve the real-time speed (30 fps). A trivial approach of fixedly reducing the forest also produced the real-time speed but with a lower accuracy ($mae = 1.60$).

10.5.3   Semiautomatic training

We used the Train Station dataset [12] for evaluating the performance of the semiautomatic training technique in Section 10.4.3. This dataset recorded over 100 persons traversing inside a train station. Sample-synthesized training images are shown in Figure 10.13.

The mean deviation errors of counting results with different numbers of synthesized images are shown in Figure 10.14. The best result by manual training using the same number of training real images is also shown in the same graph. From the graph, we can observe that the counting performance is improved with an increasing number M of synthesized training images and even get closed to the performance of the manual training if M is large enough.

Table 10.4   Detailed evaluation on the UCSD dataset with the "max" split setting (160 training, 1,200 testing frames)

Method	Error (*mae*)	Runtime (ms)	Dictionary size (MB)
Baseline [11]	2.10	82	44.9
CF with one split layer	1.89	92	1.7
CF	1.59	96	**1.5**
CF+prior	**1.43**	98	2.9
CF+prior+speedup	1.54	**36**	2.9

CF, COUNT forest; prior, crowdedness prior, speedup: forest permutation.

[†]We used only the upper split layer with a larger maximum height 10 and stored a densely structured label in each leaf node expressing the averaged density patch calculated from the original vector labels.

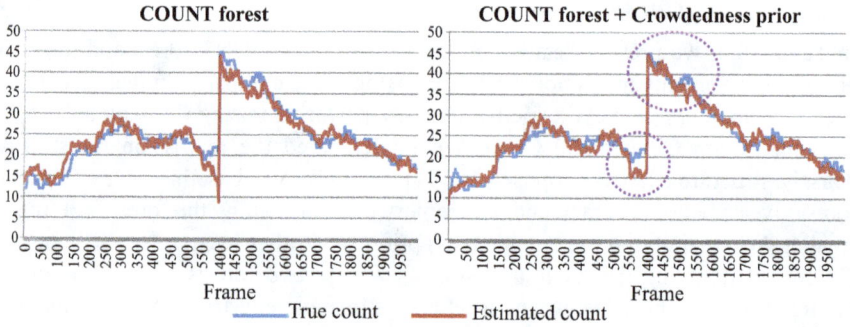

Figure 10.12 *Counting results on the UCSD dataset using COUNT forest without and with the crowdedness prior. The right graph shows better estimation in extremely crowded and non-crowded scenes (highlighted regions)*

Figure 10.13 *Real images (left) and synthesized images (right) from the Train Station dataset [12] (a) and UCSD dataset [9] (b)*

10.5.4 Application 1: traffic count

We apply our method to solve the problem of simultaneous density estimation of pedestrians and cars in traffic scenes. It has many potential applications such as accident detection and traffic control. The main difficulty is that there are noises in

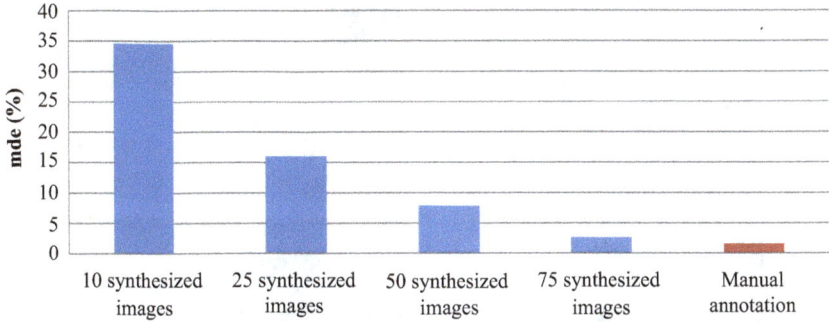

Figure 10.14   Counting results of training by manual annotation, and by the synthesizing method with different numbers of synthesized images. In this experiment, training with 75 synthesized images converged to the minimum mde (same result for 100 images)

person density estimation caused by misclassification of cars, because patch features of persons are similar to that of some car parts.

Our solution starts by making a copy of the input image for each person and car class, which we call the class layer. We then scale the car layer down by 4×4 times to match the car size to the person size. The advantage of this preprocessing is that it can filter out person appearances from the car layer, because person instances become so small after the resizing. As a result, noises of car density estimation caused by person misclassification can be reduced. We train different COUNT forests for each class layer and use these forests to estimate a density map for each object class. We then scale the car density map up to the original image size and score each class at each pixel with a density smoothed over the neighborhood region. For each location, we take the maximally scoring class along with the corresponding density as output.

We introduce a new intersection dataset for evaluating the performance of simultaneously counting pedestrians and cars. This video has a length of 11 min, with 30-fps frame rate, and $1{,}920 \times 1{,}080$ resolution. In this scene, cars and pedestrians traverse an intersection, making the scene become dense and sparse periodically. The number of pedestrians varies from 10 to 120, while the number of cars varies from 0 to 20. We sampled 100 frames from the video with an interval of 100 frames and used the first 50 frames for training and the remaining 50 frames for testing.

The density estimation results are shown in Figure 10.15. We can observe that car and pedestrian regions are correctly classified even when they locate close together. Some interesting results are shown in the first two rows, in which a motorcycle and two persons on a bus are correctly detected. They prove the robustness of our method. The performance comparison results of counting cars and pedestrians for this dataset are shown in Table 10.5. Compared with separated class counting, simultaneous counting improved the counting performance for both the car and pedestrian classes. We also obtained better results than the deformable

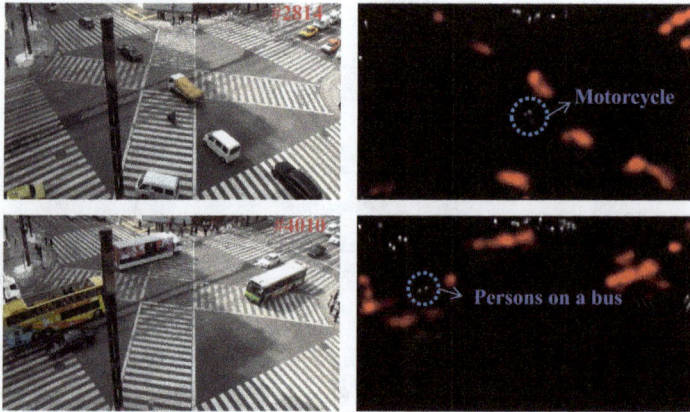

*Figure 10.15   Pedestrian and car densities estimated for the intersection dataset
(person densities are in white, car densities are in red)*

*Table 10.5   Mean absolute error (mae) of counting cars and
pedestrians for the intersection dataset*

Methods	Pedestrian	Car
DPM [31]	17.7	3.4
Separated class counting	15.4	1.6
Simultaneous counting	6.6	1.4

part models detector [31], which is the most used method for detecting multiple-class objects.

10.5.5   Application 2: stationary time

Our second application is the estimation of the stationary time, which is defined as a period that a foreground pixel exists in a local region allowing local movements [32]. The stationary time estimation can help scene understanding and provide valuable statistics computed over time. Besides, it was confirmed in [32] that simply detecting foreground at individual frames and computing how long a pixel has been in the foreground gave poor results. Our solution is to estimate a density map at each frame and compute how long a density at a pixel has been larger than a threshold. Although our method is much simpler than [32] without complex calculations, we obtained a similar result as [32]. In Figure 10.16, we showed an averaged stationary time distribution in 4 h of the Train Station dataset [12] (see Figure 10.13 (a)). As [32], it can be observed that stationary groups tend to emerge and stay long around the information booth and in front of the ticketing windows.

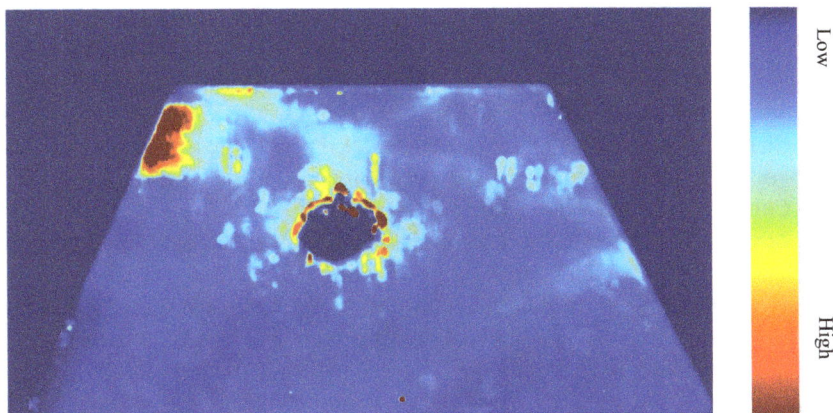

Figure 10.16 Averaged stationary time distribution over 4 h

10.6 Conclusions

This chapter presents a patch-based approach for crowd density estimation in public scenes. We learn the nonlinear mapping between patch features and relative locations of all the objects inside each image patch using a random forest framework. The patch feature used in our method is much simpler and scene independent. We build the forest in a coarse-to-fine manner with two split node layers and further propose a crowdedness prior and an effective forest reduction method for improving the estimation accuracy and speed. We achieved state-of-the-art results on the public Mall [6] and UCSD datasets [9] and confirmed high performance for multiple-class counting and scene adaptation with good results for the Train Station dataset.

References

[1] Ge W and Collins RT. Marked point processes for crowd counting. In: IEEE Proc. Comp. Vision Pattern Rec.; 2009. p. 2913–2920.

[2] Zhao T, Nevatia R, and Wu B. Segmentation and tracking of multiple humans in. Trans Pattern Anal Mach Intell. 2008;30(7):1198–1211.

[3] Brostow GJ and Cipolla R. Unsupervised Bayesian detection of independent motion. In: IEEE Proc. Comp. Vision Pattern Rec.; 2006. p. 594–601.

[4] Rabaud V and Belongie S. Counting crowded moving objects. In: IEEE Proc. Comp. Vision Pattern Rec.; 2006. p. 705–711.

[5] Chen K, Gong S, Xiang T, *et al.* Cumulative attribute space for age and crowd density estimation. In: IEEE Proc. Comp. Vision Pattern Rec.; 2013. p. 2467–2474.

[6] Chen K, Loy CC, Gong S, *et al.* Feature mining for localised crowd counting. In: Proc. British Machine Vision Conf.; 2012. p. 21.1–21.11.

[7]   Lempitsky V and Zisserman A. Learning to count objects in images. In: Proc. Advances in Neural Information Processing Systems; 2010. p. 1324–1332.
[8]   Fiaschi L, Nair R, Koethe U, *et al.* Learning to count with a regression forest and structured labels. In: Proc. Int. Conf. Pattern Rec.; 2012. p. 2685–2688.
[9]   Chan AB, Liang ZSJ, and Vasconcelos N. Privacy preserving crowd monitoring: Counting people without people models or tracking. In: IEEE Proc. Comp. Vision and Pattern Rec.; 2008. p. 1–7.
[10] Loy CC, Gong S, and Xiang T. From semi-supervised to transfer counting of crowds. In: Proc. Int. Conf. Comp. Vision; 2013. p. 2256–2263.
[11] Kontschieder P, Rota Bulo S, Pelillo M, *et al.* Structured labels in random forests for semantic labelling and object detection. Trans Pattern Anal Mach Intell. 2014;36(10):2104–2116.
[12] Zhou B, Wang X, and Tang X. Understanding collective crowd behaviors: Learning a mixture model of dynamic pedestrian-agents. In: IEEE Proc. Comp. Vision Pattern Rec.; 2012. p. 2871–2878.
[13] Pham VQ, Kozakaya T, Yamaguchi O, *et al.* COUNT forest: CO-voting uncertain number of targets using random forest for crowd density estimation. In: ICCV; 2015.
[14] Okada R. Discriminative generalized Hough transform for object detection. In: Proc. Int. Conf. Comp. Vision; 2009. p. 2000–2005.
[15] Gall J and Lempitsky V. Class-specific Hough forests for object detection. In: IEEE Proc. Comp. Vision Pattern Rec.; 2009. p. 1022–1029.
[16] Amit Y and Geman D. Shape quantization and recognition with randomized trees. Neural Comput. 1997;9(7):1545–1588.
[17] Breiman L. Random forests. Mach Learn. 2001;45(1):5–32.
[18] Gall J, Razavi N, and Van Gool L. On-line adaption of class-specific codebooks for instance tracking. In: British Machine Vision Conf.; 2010.
[19] Yao A, Gall J, and Van Gool L. A Hough transform-based voting framework for action recognition. In: IEEE Proc. Comp. Vision Pattern Rec.; 2010. p. 2061–2068.
[20] Dantone M, Gall J, Fanelli G, *et al.* Real-time facial feature detection using conditional regression forests. In: IEEE Proc. Comp. Vision and Pattern Rec.; 2012.
[21] Dantone M, Gall J, Leistner C, *et al.* Human pose estimation using body parts dependent joint regressors. In: IEEE Proc. Comp. Vision and Pattern Rec.; 2013.
[22] Ferryman J. PETS 2009 dataset; 2009. http://www.cvg.rdg.ac.uk/PETS2009/. Available from: http://www.cvg.rdg.ac.uk/PETS2009/.
[23] Rother C, Kolmogorov V, and Blake A. GrabCut: Interactive foreground extraction using iterated graph cuts. ACM Trans Graph. 2004.
[24] UCSD dataset. www.svcl.ucsd.edu/projects/peoplecnt/.
[25] Mall dataset. https://personal.ie.cuhk.edu.hk/~ccloy/files/datasets/mall_dataset.zip.
[26] Benenson R, Omran M, Hosang J, *et al.* Ten years of pedestrian detection, what have we learned? In: Proc. Eur. Conf. Comp. Vision, CVRSUAD Workshop; 2014.

[27] Van Gestel T, Suykens JAK, De Moor B, *et al.* Automatic relevance deter-
      mination for least squares support vector machine regression. In: Int. Joint
      Conf. Neural Networks; 2001. p. 2416–2421.

[28] An S, Liu W, and Venkatesh S. Face recognition using kernel ridge regres-
      sion. In: Computer Vision and Pattern Recognition; 2007. p. 1–7.

[29] Liaw A and Wiener M. Classification and Regression by random forest.
      R News. 2002;2(3):18–22.

[30] Arteta C, Lempitsky V, Noble JA, *et al.* Interactive object counting. In: Proc.
      Eur. Conf. Comp. Vision; 2014.

[31] Felzenszwalb PF, Girshick RB, McAllester D, *et al.* Object detection with
      discriminatively trained part based models. Trans Pattern Anal Mach Intell.
      2010;32(9):1627–1645.

[32] Yi S, Wang X, Lu C, *et al.* L0 regularized stationary time estimation for
      crowd group analysis. In: IEEE Proc. Comp. Vision and Pattern Rec.; 2014.

Part V

Detecting factors affecting traffic

Chapter 11

Incident detection

Neil Hoose[1,2]

11.1 Introduction

All transport networks are subject to unexpected disruption. The identification that disruption has taken place is usually given the broad title of 'incident detection'. Let us start this chapter by addressing two questions raised by that title, namely, what do we mean by 'incidents' and why do we need to detect them?

Taking the first question, we can consider an 'incident' to be an unplanned event that results in a set of consequences. These impacts might be a reduction in safety or increase in the level of hazard to road users, including maintenance and vehicle recovery workers who are present in the road; a loss of highway capacity that leads to vehicles being delayed and traffic congestion or, of increasing concern and importance, an unacceptable increase in air pollution. The latter is also a consequence for nearby residents and workers and so affects people who are not travelling. An individual incident may create all of these consequences.

To give an example, a coach broken down in a live lane on a high-speed road creates an increase in hazard for the driver and passengers as the risk of the vehicle being struck by an approaching vehicle is heightened. There is an increase in hazard for approaching vehicles because there is a stationary object in the carriageway that they must avoid while at the same time avoiding a collision with other vehicles in their immediate vicinity. The broken-down vehicle removes one lane from the road capacity and those passing in adjacent lanes will also have slowed significantly and this will reduce the practical capacity of the road. If this loss of capacity is below the current level of demand then a queue will form and start to grow upstream creating a hazard for approaching traffic, although the presence of the queue will protect the causal stationary vehicle, thereby mitigating that hazard. The presence of queuing traffic will lead to an increase in air pollution from crawling and stopped vehicles.

The incident itself may not be traffic related or even be within the highways itself, for example a grass fire where smoke is blown across the road. However, the

[1]Centre for Transport Studies, Department of Civil and Environmental Engineering, Imperial College London, London, United Kingdom
[2]Bittern Consulting Ltd, Banbury, United Kingdom

effects of the incident have an impact on nearby roads and those travelling on them. The consequences of off-highway incidents may be effectively the same as for an on-road unplanned event. Smoke blown across the road increases the risk of a collision as drivers react to much reduced visibility. This will create a localised loss of capacity from slow vehicles, or possible closure of the road leading to queues and, hence, a similar situation to our previous example of a broken-down vehicle. Non-traffic incidents will tend to be reported via different channels to traffic-related events, particularly in urban areas, so that traffic managers may not become aware of the presence of a problem until the impact on traffic is significant.

In some circumstances we can determine that the typical consequences of an incident are happening, i.e. unexpected or excessive congestion, but no actual event can be identified. Such circumstances, sometimes referred to as 'phantom' incidents, can occur where there is a loss of highway capacity but the cause is ephemeral and has gone before it can be detected, e.g. near-miss collision, stalled vehicle at a stop line that subsequently restarts.

From this discussion of the nature of incident, we can see that what we mean by an incident is very broad and what we are particularly concerned with is the consequences of the occurrence for traffic and the environment. The nature of the incident is important because that determines how the cause is resolved but the impact on traffic is also important and that means that an incident is not just the initial event but also the impacts that unfold.

Turning to the second question of why we need to detect incidents, it is clear that they create a range of problems to be resolved including mitigating additional potential hazards to drivers, passengers, responders and other roadworkers as events evolve over time. The original incident has to be responded to in an appropriate way and in an appropriate timescale. It is well known that the time between suffering a serious injury and receiving treatment can make a significant difference to the medical outcome. Quick attendance by paramedics can substantially improve the chances of survival for those with life-threatening injuries. Therefore, the time to detect that an incident has occurred is a key component. Detection is clearly a necessary pre-cursor to managing and resolving the incident.

As we have seen from our consideration of the nature of incidents the primary event is not the sole source of concern. Consequential or secondary hazards need to be recognised and steps are taken to reduce the probability of such hazards turning into further incidents. So not only do we need to detect that the causal event has taken place, but we need to be able to monitor how the situation is evolving and how the impacts are promulgating into the traffic stream.

We also need to be aware of incidents to manage and mitigate the effects on the wider network. Incidents cause a temporary loss of capacity and hence congestion on the roads approaching the site of the cause. We need to warn travellers of delays so they can choose to change their route or time of their journeys, if possible, in order to divert demand away. Techniques for accurately forecasting in real-time the spread of congestion across the road network are still in their infancy. Although they are beyond the scope of this book, the data that drives them will come from the same sources and being able to support whole-life management of an incident as well as the initial detection is a consideration in the selection and design of suitable systems and technologies.

It is quite rare that we can detect an incident directly with current technology or techniques. Mostly we infer the presence of an incident by either detecting its impact on the traffic or by identifying an unusual or unexpected behaviour of an individual vehicle. The exception to this is incidents resulting from extreme weather. The difference between weather and other types of unplanned event is that extremes can be forecast to quite a high degree and monitoring of the weather conditions locally is more likely to identify the problem sooner than waiting until the impact on the traffic becomes measurable.

Recognising this difference between weather-related events and other incident causes, and hence the approach to detection, we have divided the detection elements of the chapter into two sections; traffic parameter-based techniques and weather-related techniques. Prior to that we discuss the context of incident detection within the incident management process and identify some key indicators of performance for incident detection.

11.2   Incident detection in the context of the incident management process

The detection of an incident, i.e. an unplanned event with consequences for road users, triggers the series of steps that are needed to resolve the situation. Effective detection is clearly essential. However, the nature of that detection step influences the subsequent steps and the effectiveness of the overall process of incident management and resolution (Figure 11.1).

The setting where an incident has occurred exerts a strong influence on the detail of the incident management process and the relative priorities of different information and action. In urban streets, both non-motorised modes, such as walking and cycling, as well as road-based public transport, typically buses and trams, can be affected alongside private and freight traffic. The congestion caused by an incident can spread across the network rapidly and affect traffic that is not using the primary affected link or junction. Not only is this disruptive to travellers but it can significantly delay the arrival of the resources needed to resolve the

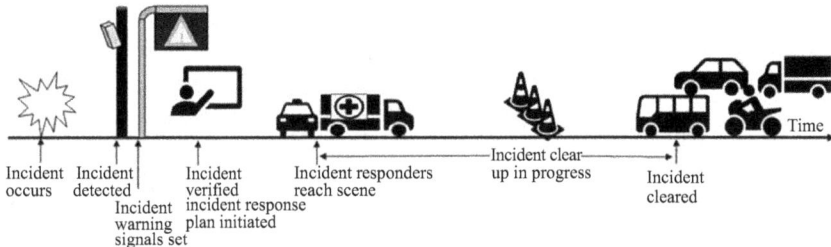

Figure 11.1   Incident management process

incident. Early detection combined with the ability to monitor the spread of congestion is key to the incident management process. Because traffic speeds are usually quite low on urban roads the safety hazards are not raised significantly. However, air quality can be very adversely affected, and this can impact on nearby residents, schools, hospitals and business premises.

On high-speed roads with speed limits of 80 kmh (50 mph) or higher, irrespective of whether they are urban, peri-urban or rural, the presence of an incident within the live traffic lanes increases the hazard of collisions for approaching traffic and raises the risk of additional, secondary incidents. Hence, mitigation of these risks is of primary importance. Rapid and accurate detection is key to getting the mitigation in place as early as possible. High-speed roads are linear in nature with few junctions spaced a few km apart so network spread takes longer, although an incident just downstream of a junction can affect the adjacent network quite quickly. Locating the incident accurately is important because the responders need to know which carriageway in order to plot the most appropriate route to get to the scene. High-traffic volumes mean that the tail of any queue can move upstream rapidly. As this is the location of the secondary hazard, the ability to detect the position of the queue tail is an important feature for detection systems on these types of roads.

Rural roads are a challenge because of the relative remoteness of any location from detection and communications infrastructure and the distance responders may have to travel. Speeds on rural roads can be high, while alignments can be poor with limited forward sight lines and hence increased risk for approaching drivers. Accurate location and a low false alarm rate may be more important than speed of detection, although the latter cannot be ignored.

Time of day is also an important contextual factor. Incidents in peak hour will have a significant and widespread effect and may take longer to resolve because responders are more delayed in reaching the scene. Also, any queues that build up will grow quickly extend farther and take longer to disperse. Incidents that occur a short while prior to the peak period starting but are not detected and resolved will tend to have the same scale of impact as peak period events. Hence, early and accurate detection that enables the incident management process to be initiated is important. During the other, less-trafficked periods of the daytime, inter-peak and overnight, the main factor is road safety and mitigating the risk to approaching traffic and those responders dealing with the scene, particularly those who arrive first. Again, early detection that enables risk mitigation measures to be put in place is important, e.g. upstream closures, warning messages on electronic signs.

11.3   Key parameters for incident detection

Several indicators that can be used to characterise incident detection technologies. Table 11.1 sets out the most significant. The actual, quantified values for these parameters are influenced by site characteristics, the wider traffic management system that the incident detection function usually sits within and the operational

Table 11.1 Key parameters for incident detection systems

Parameter	Calculation	Comment
Mean time to detect (min:s)	Average over a sample of (time of reported detection—actual time incident occurs)	This can be hard to measure as the actual time the incident occurs may be difficult to determine exactly and may not be recorded
Detection rate (%)	No. of incidents detected correctly ÷ no. of actual incidents	Usually determined over a sample of known incidents
False alarm rate (%)	No. of incorrect incident detection reports ÷ total no. of incident detection reports	Indicates the likelihood a given detection report is likely to be correct
False alarm frequency (n/time period/detection site or n/time period/carriageway km)	No. of incorrect incident detection reports in a time period for a given detection location or length of carriageway	Time period is typically 24 h. Rate per detection site is easier to measure and more relevant as the number of detection sites per unit carriageway distance may vary with horizontal and vertical alignment of road or with or presence of overbridges, gantries, and other physical characteristics
Location accuracy (m)	Distance between the location of the actual incident and that given in the incident report	In reality, this may be a fixed range determined by the location and size of the detection zone. This is of more relevance for technologies based on reports from individual vehicles or mobile devices

processes of the traffic management organisation. Therefore, when we discuss technologies, we will make qualitative comments but not provide numeric values. When determining the requirements for an incident detection system the designer should set out target ranges for these parameters based on the broader systems requirements and business case. These can then be discussed with the supply chain to determine how well any particular solution can meet these targets and any trade-offs that may be needed to reach a cost-effective solution.

The detection rate and the mean time to detect are the parameters that indicate how effective a particular technique will be. They are not unrelated as the longer the detection algorithm takes, and the more data it processes, the better the detection rate may be, up to a maximum level. However, this extended time may be too long to ensure the incident response is as good as it could be. For example, if the detection rate improves from 80% to 85% but the mean time to detect extends from 1 to 5 min, this could mean that warning signals for approaching drivers are delayed and there is a substantial increase in the risk of additional vehicles running into the primary incident.

The false alarm rate and, more particularly, the false alarm frequency are important for system credibility. As detection is just the first step in the incident management process, every alarm has to be assessed by the incident verification process, usually a manual process in a control room. Control room operators soon lose faith in a system that 'cries wolf' too often and will start to ignore it. This increases the chance of a genuine alert being missed or the response being delayed until the operator gets around to examining the report. Unfortunately, what constitutes 'too often' is a subjective view that varies between control rooms, between control room operators and with the current workload on those operators. In the case of systems that automatically set signals on the roadside to warn approaching drivers, for example the Highways England Motorway Incident and Signalling System (MIDAS), the credibility with drivers is undermined if they do not encounter any visible problem. Next time they encounter the signals, they may not heed the warning so false alerts reduce the effectiveness of signals in mitigating the risk of secondary collisions.

False alarm frequency can become an issue when a system is scaled up from a small pilot. Consider a system with four detection sites per carriageway-km where the false alarm frequency is 1/day/detection site. For a pilot trial of 5 km of highway, this would give $5 \times 2 \times 4 = 40$ false alarms per day. This is less than 2 per hour which might seem acceptable. However, if the system is rolled out to 100 km of highway managed by a single control room then the number of false alarms would increase by a factor of 20–800 per day which is more than one every 2 min! False alerts do not tend to be evenly spread but come in clusters, as a result of the triggers for false alerts also being uneven, and a control room could easily be overwhelmed by false alarms.

Detection rate and false alarm rate and frequency are related for most incident detection approaches. Increasing the detection rate typically involves increasing the sensitivity to some measurement parameters or thresholds and this increase the probability of false detections.

Therefore, a practical incident detection system must achieve a high enough detection rate to realise benefits that exceed its cost by a margin determined by the funders method of economic appraisal while keeping the false alarm frequency and false alarm rates low enough to maintain the credibility of the system in the eyes of the system operators and the travelling public. This is a judgement that can only be made in the context of the actual deployment. Care needs to be taken to explore these trade-offs when moving from a restricted trial implementation to a full-scale, business as usual deployment. Our experience is that incident detection systems need periodic review to ensure that they are tuned to maintain the expected levels of performance as this will change over time due to changes in traffic, drivers modifying their behaviour and normal deterioration in technology as it ages.

11.4   Incident detection using traffic-parameter-based technologies and techniques

It is unusual to be able to detect an incident directly. In most cases, we detect an unexpected change in a measured set of parameters and from that infer the presence of an unplanned event. We measure the impact of the event on the traffic or other vehicles, use this to raise an alarm and then use other methods, e.g. CCTV (closed-circuit television), on-road patrols, to verify the presence and nature of the cause. Incident detection is a process involving measurement from sensors feeding into methods for analysing that data to infer the presence of an incident. The focus of this chapter is on sensor technologies and how they function rather than on the algorithms used to perform incident detection based on sensor output. Readers can find a wide range of papers in the literature and reviews can be found in [1,2].

The increasing penetration of mobile devices and vehicle-infrastructure connectivity, linked to in-vehicle sensors, is changing the landscape and creating the potential to move the point of detection closer to the primary incident in both space and time. We shall discuss these developments later in the section but before that, we shall review the techniques for incident detection based on roadside sensors and data analysis algorithms. First, we will identify the parameters of traffic that are typically measured, then we consider the sensor technologies available to collect that data.

11.4.1   Flow in vehicles per hour per lane or per direction

This represents the volume of traffic. A sudden drop in volume can indicate a blocked lane or carriageway upstream of the detector site or congestion from a downstream incident reaching the measurement location. A sudden increase can indicate a problem on a parallel route. The value is often scaled up from a count of vehicles in a shorter time interval, usually 1 min. Because vehicle flow is very random at the microscopic scale, it can be difficult to differentiate between a change because of an incident and the normal random variation unless the degree of saturation, i.e. the ratio of current volume to available capacity, is high or data is analysed of over several minutes.

11.4.2   Average speed per time interval at a specific location

This is normally given in kilometres per hour (kmh) or miles per hour (mph) per lane or direction and shows how free flowing the traffic is. On high-speed roads, reductions in speed are a clear indication of problems, although this may be caused by congestion from demand exceeding capacity as well as congestion resulting from an incident. In urban areas, speeds are lower and variable so discriminating between incidents and normal traffic conditions is more difficult.

11.4.3   Average speed over a distance, or journey time, per time interval

Journey time is the inverse of average speed over a distance and changes in magnitude can be easier to determine as a result. These values are measured over a sample of vehicles reaching the downstream measurement point and can be slow to react if that sample takes longer to accrue as a result of an incident. Journey time can also be measured from tracking an individual vehicle or mobile device using either GPS or cell-phone tracking.

11.4.4   Headway (time) in seconds average per lane per time interval

Queuing traffic will be closer together. This measure when combined with average speed can differentiate between light and dense traffic conditions.

11.4.5   Detector occupancy

This is the percentage of time in 'detect' state per unit time or the continuous duration in seconds the detector is in detect state. It only requires a single detector so is very simple to measure. The value is related to the speed of the traffic, vehicle lengths and the dimension of the detection zone in the direction of travel. High values of detector occupancy indicate slow or stationary vehicles across the detector.

11.5   Sensor technologies

11.5.1   Inductive loops

The most widespread detector technology used for traffic applications is the inductive, or magnetic, loop. The principle of operation is based on measuring the change in inductance in an oscillating current cause by presence of magnetic material in the currents' magnetic field. The induced current in the target object is in the opposite phase to that in the detector and suppresses the current in the detector by a small but measurable amount. Further details can be found in [3].

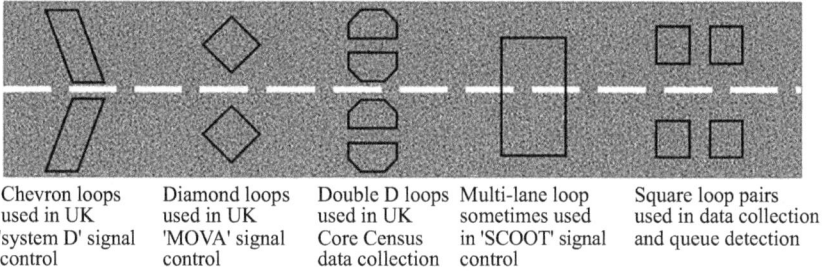

| Chevron loops used in UK 'system D' signal control | Diamond loops used in UK 'MOVA' signal control | Double D loops used in UK Core Census data collection | Multi-lane loop sometimes used in 'SCOOT' signal control | Square loop pairs used in data collection and queue detection |

Figure 11.2   Examples of different inductive loop shapes

There are many variations in loop shape depending on the main application (Figure 11.2). Where the main application is simple presence, i.e. 'vehicle' or 'no vehicle', then the loop's shape can be optimised for sensitivity. This is done by maximising the number of magnetic 'field lines' that moving objects will cut through as they pass over the loop, so diamond, chevron and circular shapes are all possible. Once we need to measure more parameters such as duration of detection, or loop 'occupancy', then we need to move to a rectangular shape where the first and last line of detection is orthogonal to the direction of travel. We also need to provide loops in each traffic lane so that we can discriminate the differences between them and get more detail on the traffic behaviour.

Rectangular loops are typically 2 m×2 m (6 ft×6 ft). The dimension across the lane is a balance between being large enough to prevent vehicles passing by the side of the loop and not being detected, and small enough so as not to detect vehicles in the adjacent lane or suffer electromagnetic interference from loops in the next lane. The loop comprises a number of 'turns' of copper cable in the road connected to a feeder cable that leads back to the source of the oscillating current and the detection electronics located in a cabinet at the roadside. The feeder cable is usually shielded in some way because the detection 'antenna' comprises both the feeder and the loop cables in the road. Feeder lengths up to 1,000 m are not unknown and 200 m is relatively common. Thus, the loop component of the detector may be a relatively small part of the overall circuit and this is why there are multiple turns of cable in the section in the traffic lane, typically 3 or 4. If we have a detector that is 2 m×2 m with 3 turns and 100 m feeder the total antenna length is $(2×100)+(3 (2+2+2+2))=224$ m but the detector is only 24 m or just under 11% of the total. Adding an extra turn in the traffic lane increases that to 13.7%, thereby improving the sensitivity of the detector because it is a greater proportion of the circuit. Obviously, reducing the feeder lengths will also improve sensitivity for a given loop size and number of turns, but the location of the cabinet to house the electronics will often be subject to other factors, e.g. safe maintenance access, that may limit the scope to do this.

Where speed data is required, a pair of loops is usually installed from a known distance apart in each traffic lane. The time that the detector changes from 'no detection state' to a 'detect' state is recorded for each loop, and the time difference between this change at successive loops can be divided into the distance between the leading edges of the loops to give a value for the vehicle's speed. Once the speed is known, this can be used in conjunction with the duration of the 'detect' state of one of the pair of loops and an estimate of the length of the detection zone to estimate the 'magnetic' length of the vehicle. This is usually slightly shorter than the true length, typically 50–150 mm (2–6 in.) because of the overhang from plastic fenders but is adequate to be able to provide vehicle classification based on length.

It is possible using signal processing techniques on the change in inductance from an individual loop to obtain an estimate of speed without the need for a second loop as discussed in [4].

Loops have a well-defined detection zone slightly larger than the area of the visible slots cut to install them. This makes it relatively easy to tune them as the vehicle being detected can be related to the output of the loop. At setup, a loop-detection array requires some tuning to ensure the sensitivity is sufficient to detect vehicles consistently without creating interference between nearby loops or detection of vehicles in an adjacent lane. Most loop electronics incorporate self-tuning to adjust automatically to changes in background conditions in the circuit and may provide some self-diagnostics that can identify if a detector is not operating correctly. Electronics are available that will control and measure from a single loop up to arrays of 32 loops.

The most significant drawback of indicative loop detectors is that the cables are vulnerable to damage. This may occur during installation where a sharp edge, voids in the backfill or unintended tears in the cable insulation can result in the copper code becoming exposed to moisture. This will result in the circuit not working correctly and either fail to detect or create random phantom detections. Thermal movement of pavement can have a similar effect on detector performance because of chafing of the cables. Any pavement works that cut into to the road, e.g. scraping off prior to re-surfacing, can lead to physical damage to the detector. Similarly, feeder cables are vulnerable to verge side disruption from utilities digging trenches. This is a particular problem in urban areas.

Reinstatement requires complete replacement of all the loops and feeders as the relative positions and continuity of cabling are important. This can require extensive traffic management. Cutting loops can affect the integrity of road structure and re-cutting loops in the same location increase the potential for early life failure of the road surface.

A carefully installed loop array that is undisturbed by external work can provide consistently good quality output for as long as the road surface lasts. Periodic checks of the tuning of the electronics should be carried out but this is probably only needed annually if self-diagnostics and remote fault reporting are available.

11.5.2   *Fixed-beam RADAR*

The radio detection and ranging (RADAR) technology was first applied to traffic sensing during the 1980s when advances in solid-state technology allowed compact and affordable devices to be made suitable for detecting the presence of traffic at signal junctions. These devices use 'continuous wave' (CW) radar to detect the shift in frequency between an emitted signal at a known frequency and the reflection of that signal from a moving object, a phenomenon known as Doppler shift. The technology can be calibrated so that the device can measure the speed of the object based on the magnitude of the shift. The direction of the shift can be used to discriminate between objects moving towards or away from the detector. Calibrated CW radars are used in some speed enforcement systems.

The limitation with CW radars is that an object will not be detected unless it is moving and the distance to the target object cannot be determined. These problems can be overcome by using 'modulated' radar beams. Instead of using a CW at a single fixed frequency, the emitted signal frequency can be changed continuously according to a predetermined configuration. This is called frequency-modulated CW (FMCW) radar as shown in Figure 11.3. The technique compares the frequency of a returned signal with the frequency being transmitted at that time. Because the rate of change of the emitted frequency is known, the difference between it and the return signal is proportional to the time taken for the signal to reach the target and return. The speed of the signal is the speed of light and hence the range to the target can be determined from:

$$\text{Target range} = \frac{1}{2}\frac{(F_2 - F_1)c}{R} \tag{11.1}$$

where F_1 is the return signal frequency, F_2 is the transmit frequency at time return frequency F_1 is detected, R is the rate of change of transmitted frequency and c is the speed of light.

Division by two is needed because time of flight is to target and back.

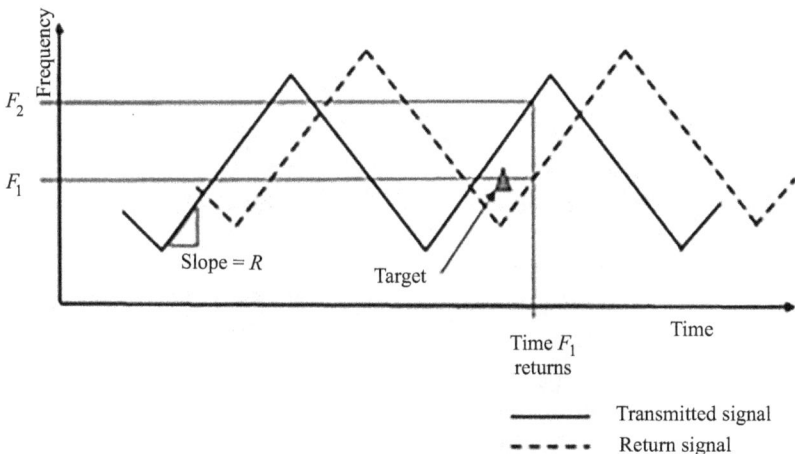

Figure 11.3 Frequency–time graph for FMCW radar

RADAR uses the microwave part of the electromagnetic spectrum and FMCW devices typically operate in the 18–26-GHz (also known as K-band) region. An individual device will operate in a narrow segment of this, for example the Wavetronix HD device operates in the range 24.0–24.25 GHz.

Doppler (CW) radars need to be mounted so that the beam is along the traffic stream as it is the movement of the vehicles that is important. If such a radar is mounted looking across the traffic, it will not be able to detect the movement component along the radar beam will be very small. For FMCW radars, this is not the case as targets can be identified whether or not they are moving and so the device can face in any direction. Furthermore, because the range of the target is known, those targets at similar distances can be grouped together and a set of detection 'zones' can be set up. Thus, an FMCW radar mounted at the side of the road with its detection beam at right angles to the traffic can have a detection zone for each lane and this single device can monitor multiple lanes of traffic. As the radar beam is orthogonal to the traffic the Doppler effect cannot be used to monitor speed. If two beams are placed in parallel a known distance apart, the speed can be determined by the arrival time of a target in each beam. This is analogous to the method used to measure speed by inductive loops.

A typical 'side-fire' arrangement is shown in Figure 11.4. A device is mounted 4–6 m above the ground and a similar distance back from road edge looking orthogonal to the direction of the traffic. The advances in radar electronics mean that the pair of radar beams can be enclosed in a single, compact housing easily mounted on a lighting column or its own pole. A single such device can monitor multiple lanes, enough to monitor both carriageways of a dual 4- or 5-lane highway from one side.

The data output from the device is the same traffic data produced by inductive loops. Therefore, this data can be fed directly into most incident and queue detection algorithms that are based on such data analysis.

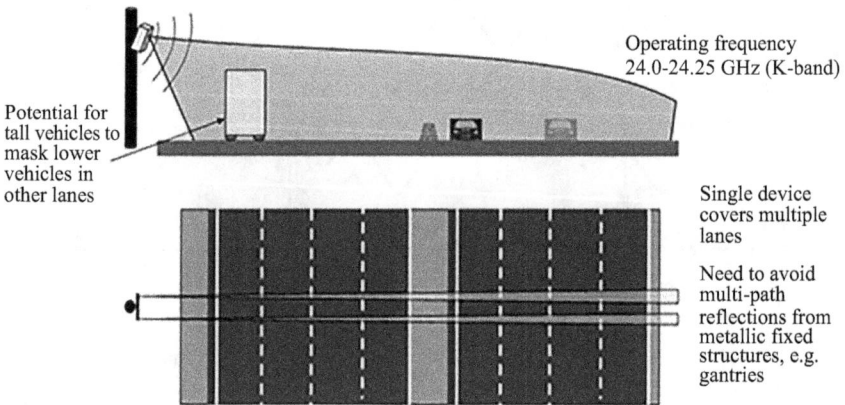

Figure 11.4   Typical arrangement for 'side-fire' FMCW radar

Radars are very tolerant of poor weather and are only affected by extremely heavy rain or snow. They are vulnerable to electromagnetic interference from other devices transmitting in the same frequency range. The quality of the data can be affected by large vehicles nearer to the device obscuring smaller vehicles further away. This problem becomes more severe with increased traffic density and with increased flows of tall heavy goods vehicles or buses in the lanes nearest the device. Radars are also subject to multi-path reflections where the radar has reflected from multiple surfaces. These can create false targets in detection zones.

Radars can be difficult to align and vibration or thermal effects in the mounting can shift the beam direction such that detection zones temporarily include fixed objects (e.g. safety barriers) that create false detections for as long as the beam is out of alignment.

11.5.3   Computer vision

CCTV is widely deployed on highway and urban road networks. Incident detection by operators watching screens containing multiple cameras images, perhaps with scrolling between cameras, is possible but it is not efficient. Control room operators are better employed managing any issues present on the network. Also, it is quite difficult for anyone to maintain a reasonable level of concentration for very long when faced with screens of multiple images so there needs to be a team of operators available to prevent any one individual having to spend more than 1–2 h watching the screens. That said, in urban areas where cameras are also being used for security reasons this technique is quite common. Such systems tend to have cameras covering areas where pedestrians are dominant, e.g. rail and bus stations, pedestrian shopping streets, and so do not have views that are particularly useful for traffic incident detection.

The availability of digital images, either by conversion of analogue images using and analogue–digital (A/D) converter or, as most cameras are now digital devices themselves, direct access to the image memory on the camera, means that the image data can be manipulated automatically using computer algorithms. The data from an image comprises one or more two-dimensional arrays of numbers. Each number, referred to as a 'pixel', short for 'picture element' represents the brightness of the light reaching the camera from that point in the real world. Monochrome images show only the total brightness and comprise a single array of pixel values. Colour images are formed from three arrays of pixel values from the brightness of the red, green and blue parts of the spectrum. The number of pixels is determined by the camera definition typically ranging from 0.25 million for a 512×512 array up to 8 million for high-definition (HD) camera systems. CCTV systems capture 25 or 30 frames per second (fps) depending on the regions mains frequency, although with digital images this is really a hangover from earlier analogue systems. In order for motion to appear smooth to a viewer, a frame rate of more than 25 fps is needed. Computer processing of data does not have the same need so a lower frame rate can be used. Also, the combination of resolution and frame rate determines the communications bandwidth needed to communicate image data streams. Limitations on communications bandwidth may restrict the amount of image data available for processing CCTV images at a location remote from the camera while providing adequate computer processing power at every camera will raise the system cost.

There are features of digital images that make this an attractive technology. CCTV images cover a wide area. The extent of the area covers depends on the lens on the camera and may range from an area of a few square metres for a narrow focus to up to 500 m continuous length of highway. The area being viewed may be near the camera's location or it may be some distance away if a zoom lens is being used. This gives great flexibility in positioning of the cameras relative to the area of road that is to be monitored and means that a single device can monitor multiple lanes of traffic in multiple directions. Because humans can view the same images as the computer it is easier for the control room to interpret output from computer algorithms. This aspect can also make setting up the system simpler.

There are two different approaches in how image data is then used. Some methods use the data from the whole of the image, whereas others only use a subset of pixel data. The former, which we term 'wide-area computer vision', is described in Section 11.6.1. In this section, we will focus on those techniques that use a subset of the available pixel data.

In most images of highways, the road occupies the minority of the image. Therefore, it makes sense to restrict data processing to those pixels that are within the highway or the vehicle trajectories. Processing of these pixels produces output that is relevant to traffic by their context. Bearing in mind that images are being produced at up to 25 fps, equivalent to 200 Mbytes/s for a monochrome image. Because the brightness value of each pixel is represented by a single data byte, reducing the number of pixels being processed allows faster processing overall or allows more complex algorithms to be run in the same time interval.

Algorithms may enhance the images for humans to view or may manipulate the data to identify and extract features that are of particular relevance and interest. In the latter case, the areas of interest in the image are pre-selected in software and can be thought of as creating 'pseudo-detectors' within the image. These pseudo-detectors can be laid out to emulate inductive loop detectors and the presence of vehicles within the image at these locations can be processed in an equivalent way. The output will then be traffic data, e.g. presence time, speed, traffic counts, for each of the pseudo-detectors in the image. Because the pseudo-detectors are created in software there can be any number of them, of any size and location subject to the amount of processing power available to meet the need to process the pixel data quickly enough. This sort of data can then be pushed to a generic traffic data incident detection of the type described in Section 11.4.2 in the same way as inductive loops, side fire radar, etc.

There are two basic forms of algorithm that underpin the computer image processing used in this type of system; background comparison and inter-image comparison. Background comparison does what it says; it compares the current image to a reference image that does not have traffic within it. This will highlight those areas within the current image where there is a difference with the reference image. By context, these will be vehicles. The reference image should not contain any vehicles at all so this technique will detect presence of both stationary and moving traffic. Inter-image comparison identifies those pixels whose brightness values have changed in the time interval between the two images and identifies movement. This information can be combined with the analysis of the background comparison to discriminate between moving and stationary vehicles.

The extraction of vehicle data will involve additional steps to group adjacent pixels that are identified as candidates from the two comparisons into single objects that represent vehicles. Further steps will measure the magnitude of any movement related to a vehicle's speed and direction of travel within the image. The size of coherent groups of pixels can be used to estimate vehicle size and classify them accordingly.

The most favourable conditions for computer vision are daylight on a cloudy day with light winds. Under these conditions then detection performance can rival those of well-tuned inductive loops or radar. However, because vision operates in the visible part of the electromagnetic spectrum it is subject to a wide variability in conditions. Any form of weather that reduces visibility, e.g. fog, rain, snow, will reduce the detection range. Strong sunlight creates shadows that can make keeping a reference image valid difficult and can create moving artefacts within the image that are not easy to discriminate from objects of interest. Highly reflective surfaces such as the roof of a vehicle trailer, or a bus roof, can affect the dynamic range of brightness within the camera and render darker, low contrast objects invisible for a short time. Wet surfaces will reflect the lights of vehicles and this can create artefacts in the image that can be mistaken for moving vehicles. Likewise, at dusk, during the night and at dawn, vehicle lights may be the dominant feature making it difficult for vehicle objects to be correctly segmented from the background. All these tend to result in a variation in performance when they are present and this can lead to significant numbers of false alarms in linked incident detection algorithms.

Changes in image geometry as a result of vibration or wind sway of the camera mounting can degrade performance of computer vision systems. The position of the pseudo-detector is fixed in the camera imaging device, but this can move relative to the ground plane if the camera moves. As a result, the pseudo-detector may move out of position on the road such that any reference image is no longer valid, or the result of inter-image comparison is dominated by the camera movement. In the same vein, it is very difficult to interface computer vision techniques with moveable CCTV camera where an operator can change the view by altering the pan, tilt or zoom (PTZ) of the camera head. Computer vision systems so far need to have a fixed view so that the pseudo-detectors can be placed. If the camera PTZ is changed then they will no longer be in position. Most PTZ CCTV systems have the ability to remember a number of pre-set camera views that can be selected by entering a code at the operators' terminal. Even in these cases the tolerances in the mechanical components mean that the pre-set position will not be identical when it is selected on different occasions and the magnitude of these differences, while irrelevant to a human viewer, are sufficient to place some or all of the pseudo-detectors out of position.

Any highway manager implementing this technology needs to ensure all the aspects that degrade performance can be addressed. Locations where such problems are minimised include road tunnels where lighting is controlled, weather does not encroach, and camera positions are not subject to wind and temperature effects, although vibration may be an issue.

11.5.4   Journey time measurement using licence plates

A particular application of computer vision is to read the licence plate of a vehicle using automatic licence plate reading (ALPR), or in the UK automatic number plate reading, algorithms. Because each vehicle has a unique combination of letters and numbers if the same vehicle plate is read at two different locations at different times, then the time taken to travel between these two points, i.e. the journey time, can be determined. By continually recording journey times statistical patterns can be generated and deviations from those patterns can be used to identify potential incidents. This may be as simple as the journey time averaged over the last minute, or the last *n* recorded journey times falling below a pre-determined threshold, or it may be a more sophisticated approach comparing current values to a historic trend.

ALPR makes use of a solid-state, charge-coupled device or complementary metal–oxide semiconductor image sensor. The field of view is relatively narrow and normally has a view of 25–30 m of a single traffic lane 20–50 from the camera position. This is to ensure that a licence plate occupies enough of the image to allow reliable segmentation of the plate's region in the image and that the characters are sufficiently large and clear within the image for optical character recognition to be effective. The field of view is nearly always illuminated by additional infrared light. This allows the plate to be legible in the image even if the vehicle lights are on or the Sun is reflecting off the plate. It also improves the contrast between the characters and the plate under all light conditions, thereby improving the success of character reads.

Typically, 90% of licence plates are sufficiently legible to be read automatically because they are not damaged or partially obscured by dirt. Most systems on the market will read 80%–90% of the legible plates in the traffic lane being monitored giving a sample of 70%–80% of the traffic. The number of matched plates that create journey times depends on the amount of traffic that can take a different route and not pass the downstream camera. In urban areas, there can also be a problem with vehicles such as delivery vehicles and taxis that pass both cameras but take a circuitous route between them. This results in a very wide range of valid journey times even when there is no traffic disruption.

Another limitation with algorithms based on journey times is the time to detect can be longer than is needed to respond to an incident within an acceptable time frame. Sufficient number of plate matches of vehicles whose journey time has been increased by the incident must have reached the downstream camera or, in the case where the incident completely blocks the road in that direction, sufficient time with no readings must have elapsed to be confident the traffic flow has been disrupted.

ALPR systems are not low maintenance. Cameras lenses must be cleaned on a regular basis, particularly if the camera is subject to salt spray from passing traffic. Cleaning can be needed as often as once per month although three of four times per year is more usual. The infrared illuminator, usually an LED-based device, also has a finite lifespan and will need replacing after two or three years.

11.5.5 Journey time measurement using Bluetooth and Wi-Fi

Many vehicles are now equipped with on-board Bluetooth and Wi-Fi devices to support hands-free calls and on-board infotainment. Many vehicles have at least one smartphone on board also equipped with these communications channels and with the Bluetooth and Wi-Fi channels switched on. Both these techniques provide bi-directional wireless communications between specific devices. They differ in their frequency of operation, bandwidth and range but are common in that the initialisation of communications requires an initial request and response handshake between the two devices.

A roadside device can be installed that constantly seeks out other devices that enter within range. When a mobile device comes in range and receives the request to connect the response message it sends will include the media access control (MAC) address. This is a unique hardware identifier and can be considered as the equivalent to a vehicle licence plate. Hence, they can be matched at different roadside devices and the time of travel between devices calculated exactly as for ALPR systems.

The limitations are the same as for ALPR systems in respect of matching and time to detect. Wi-Fi has a range that can be up to 250 m from the roadside device meaning that the location of the mobile device may be too imprecise for accurate journey time measurement. There are concerns about privacy relating to use of the MAC address to track individuals. The use of dynamically allocated MAC addresses which prevents tracking is increasing so it may that this technique will not be possible at some point in the future.

The roadside devices are cheaper than ALPR cameras and have more flexibility in where they can be mounted. They are also very low maintenance as there are no lenses to clean or infrared illuminators to replace.

11.6 Wide-area incident detection techniques

In the previous section, we have considered technologies that measure over a few metres at a specific location on the road network and then pass the measurements to an incident detection algorithm which is to a large degree independent of the sensor technology. In this section, we describe technologies that sense continuously over a distance of tens or hundreds of metres. Because of the specific techniques, the method of incident detection tends to be closely coupled to the sensor technology.

11.6.1 Computer vision

A key attraction of computer vision technology is that it provides a continuous image of the road that can be several hundred metres long. However, the practical range is usually around 150–250 m because of considerations such as ensuring the target objects are big enough within the image to be reliably identified, curvature of the road, the amount by which large vehicles near the camera obscure smaller ones further away and weather conditions.

There are two broad approaches to processing the image data. The first is to use a combination of segmentation techniques including reference image comparison, inter-frame differences and feature extraction operators on the current image. The latter are mathematical algorithms applied to small regions of the image, typically 3×3 or 5×5 pixel areas, which group together regions with similar intensities or highlight where the image intensity changes to identify the edges of possible objects. Adjacent pixels with common features are grouped together to form target objects and these targets are then matched across successive images so that their trajectory in time and space can be traced. Unusual or unexpected trajectories can be used as an indicator of the presence of an incident.

One of the challenges of this approach is to be able to convert the trajectory as extracted from the image into a trajectory on the ground. The camera does not have a bird's eye plan view. Feasible camera mounting heights are in the range 8–12 m in most locations, so the camera is looking along the road at a shallow angle in order to achieve the range of view. Each pixel in the image has the same internal angle of view but the area on the ground will vary substantially. Pixels at the lower edge of the image will each cover a small area within the foreground, while pixels at the upper edge each represent a larger area in the distance. The size of these areas is a geometric function of the lens angle, pixel density in the imaging device, camera height and camera tilt. It is not possible to extract distance information directly from a monocular camera without some ground reference points at known locations so there must be some clearly visible objects in the image that can be used to provide calibration data.

Stereoscopic vision that uses a pair of cameras set close together and looking in the same direction to overcome the calibration issue as depth data can be calculated from a knowledge of the camera characteristics and the distance between the cameras. The technique has been used for vehicle measurement and classification at toll plazas, but the additional cost of the equipment may render them unviable for the business case for incident detection applications.

The second approach is to use machine learning techniques on the image, or on the results of segmentation of the image, to identify possible incident conditions directly without explicitly extracting vehicle or target data. Possible techniques are neural networks, which mimic the way neurons in an animal brain work, or k nearest neighbours, which classify which set the data under examination is most likely to belong to. These methods make use of a 'training set' of data, i.e. data where we know the outcome, to establish their internal weights and parameters. Once a stable set of parameters has been established these are used to process incoming data and classify it accordingly. The advantage of this approach is that we don't have to describe in detail what an incident or congestion looks like, we just need some example image sets we can use for training. We can improve the resilience to weather or lighting variations if we can include examples of these variants within the training set.

Wide area computer vision has the same limitations as described in Section 11.5.3. Machine learning methods may be more tolerant than algorithmic approaches, but they are sensitive to the training set. This may result in a lack of sensitivity, e.g. missing genuine incidents, or too many false alarms, or it may even swing between the two under different conditions.

11.6.2   Scanning radar

We have already described in Section 11.5.2 how a narrow radar beam pointing across a highway can be used to detect the presence of vehicles and that the range of the vehicle from the detector can be measured and used to determine which lane it is in. If, instead of being fixed, the radar detector is mounted on a rotating platform then it can scan through 360° and report the range of all the reflecting objects it detects. A typical arrangement is for the device to be mounted on a pole at the side of the road about 6 m above ground level and for the radar to fully rotate at 4 or 5 rotations per second.

A reference 'radar' image can be created of all the static reflections and, by comparing each new scan with this reference, moving objects can be identified and subsequently tracked from scan to scan. A software 'mask' can be applied that limits the area of interest to those targets that are within the highway boundary so that the system does not attempt to track animals in adjacent fields or traffic on other roads. Target trajectories can then be tracked and an alert raised if the characteristics fall outside the expected range, e.g. the speed falls below a certain threshold, the vehicle path suddenly changes.

This technology has been deployed for highway and tunnel incident detection in Sweden, tunnel incident detection in the United Kingdom and for isolated stopped vehicle detection on all-lane running sections of the Highways England 'smart' motorways in the United Kingdom. Detection performance is good with a short time to detect, a high detection rate and tolerable false alarm rate. The radars are unaffected by weather except where strong winds create vibration in the mounting or cause vegetation to sway into the detection zone. The range of the device is up to 500 m radius from the device although the practical range is more likely to be 250–350 m even on a straight highway. A single device can cover both sides of a dual carriageway road with any number of lanes.

There are several factors that can affect the performance of a scanning radar. The mounting needs to be stiff and prevent sway or vibration. This is because a relatively small movement of the radar head can result in a significant change in the position of the radar beam at ranges of more than 200 m. This can result in the current image not mapping well with the reference image and false artefacts being created. Also, the software mask may not be valid at this time and objects outside the field of interest will be identified and tracked inadvertently. Fixed items along the road, notably bridge piers, gantry legs and lighting columns will cast a radar 'shadow' and the system will not be able to detect anything inside that shadow thus limiting the range. At the more extreme of the range, the spacing of lighting columns becomes important as the shadows of successive columns may overlap creating a continuous shadow. If the columns are in the central reservation, then the carriageway opposite to where the radar is positioned may be effectively invisible. Similarly, tall vehicles stopped near the radar will cast a shadow and this may go into the carriageway depending on the relative positions. Care should be taken to consider the location of any parking or emergency bays relative to the radar position as it may seem attractive to position the device near such bays for ease of maintenance access.

Radar is reflected by all sorts of surfaces including organic materials such as plants or animals. The growth of vegetation in the spring can lead to additional artefacts that need to be incorporated into the reference image. A more difficult problem is vegetation on the edge of the mask that becomes detected inside the area of interest when the wind blows strongly enough in a particular direction. This can create random false alarms unless a higher level processing step is incorporated to identify and suppress the tracking of this type of target.

At its farthest range, the horizontal resolution of the radar beam covers a significant width and the device cannot resolve which lane a target is in and may not be able to resolve which carriageway the target is on. This makes verifying the location of the incident by associated CCTV more difficult and is another reason why the practical range is shorter than the range at which detection is possible. Finally, the radar detection zone generally only starts 15–20 m from the detector when the radar beam intersects the ground. This means there is a 'blind spot' in the immediate vicinity of the radar position.

11.6.3   Use of linear radar

It is possible to monitor a single lane of traffic by aligning an FMCW radar beam to look along a lane of traffic rather than orthogonal to it. The narrowness of the radar beam means that only a single lane can be monitored but the device can detect and track vehicles up to 200 m. Instead of creating pseudo-detection zones the system detects and tracks vehicle targets along the whole radar beam. The behaviour of the targets, notably their speed, can be used to infer the presence of an incident or the congestion caused by one.

11.6.4   Light detection and ranging

Light detection and ranging (LiDAR) uses pulse encoded laser light to measure the distance from the emitting device to any reflective surface. Unlike radar, the beam is very narrow in both vertical and horizontal directions. Devices usually comprise a number of light emission-detection pairs mounted vertically. These can either be fixed or mounted on a rotating platform to provide a swept area of detection. The output from the sensing is a referred to as a 'point cloud', which is a set of data points where each point is defined by a vector of horizontal angle, azimuth, range and reflected intensity. These devices have many applications including traffic measurement and are perhaps best known as part for their use in Google Streetview and Google-automated vehicles. They have been used for highly accurate count and classification in toll booths and, in principle, could be deployed in a manner similar to scanning radar to provide target tracking and incident detection. However, they are more expensive than scanning radars with no specific additional capability or performance.

At the time of writing a new form of LiDAR called 'flash', LidAR is being used in aerial surveying. This emits a continuous burst of laser light across a wide angle in both planes. Reflected laser light is detected by a solid-state array detector analogous to the detectors used for CCTV cameras although with much lower resolution. This has the potential to create much denser point clouds over a well-defined area of interest more equivalent to a camera field of view. Hence, once the costs become lower, a new option may be available that combines the benefits of camera like views and LiDAR/radar ranging.

11.6.5   Longitudinal optic fibre

Optic fibres are primarily used for high bandwidth digital communication but they can also be used as highly sensitive vibration sensors. Small deformations in a fibre caused by nearby vibrations can be detected as changes in a known optical signal along the fibre. Originally developed for the defence and security sectors, signal processing applied to a specific signal carried down a single fibre can be used to determine the presence and linear location along the fibre of a vibration in the surrounding ground. The system is sensitive enough to detect footsteps at a range of several kilometres along the fibre from the signal processing equipment. The distance away from the fibre that a vibration can be detected depends on the magnitude of the vibration but vehicles can be reliably detected up to 20 m away.

For a fibre running alongside or underneath a road, both the time and location of the vibrations from individual vehicles can be recorded. This creates a time-distance graph for all the vehicles all the way along the fibre, potentially a distance of several kilometres. Unusual or unexpected patterns in the time–distance data can be used to infer the presence of an incident and specify its location to within a few metres. Additionally, the system can detect the presence of pedestrians in the carriageway and this may also be an indicator of an incident.

The technique has clear potential on long stretches of highway where there a few other sources of vibration. The technique only needs a single fibre from the large bundle of fibres in an optic fibre cable. Hence it can be applied using an unused fibre (sometimes called a 'dark' fibre) in an existing communications cable. The signal processing equipment does not need to be at the roadside and can be housed in a communication building at one end of the fibre along with other fibre optic transmission equipment. This makes maintenance of the system much easier.

This technology has some limitations. It requires that the fibre follows the alignment of the highway sufficiently close so as not to have blank regions where the lateral distance between traffic and fibre is too far for vibration to be detected. For a dedicated cable, this should not be an issue but one of the attractions of this method is to utilise dark fibres in existing optic fibre communications installations, and in these cases, the alignment may not be suitable. The method is not suitable where other sources of vibration, for example an adjacent railway line, might effectively drown out the vibrations from traffic. As a consequence, it is difficult to see that it can be applied to urban streets where patterns of vibration will be much more complex than for rural highways. There is also a resilience issue because if the optic fibre is damaged or cut this will potentially remove detection from the whole length of fibre, and certainly for that length beyond the damage away from the signal processing equipment.

11.6.6   *Mobile phone, probe vehicle and connected-autonomous-vehicle-based techniques*

So far we have considered techniques that use a measurement technology that is fixed in location and largely independent of the vehicle, excepting the Bluetooth and Wi-Fi techniques discussed in Section 11.5.5. Pervasive, wide-area wireless data communications provided by 4G mobile phone networks means that handsets and suitably equipped vehicles can be tracked. This gives us the potential to identify unexpected patterns in movement from a rich source of data that is available anywhere on the network not just where monitoring devices have been installed. Much of this information is a by-product of provision of other services such as mobile internet, dynamic navigation and commercial vehicle fleet management.

Mobile phone networks need to track handsets in order to be able to route calls and data. This means the network operator knows, as a minimum, the location to the nearest cell in the network of any powered-up device equipped with a valid SIM card. The strength of the signal can be used to give an estimate of the range of the mobile device from the fixed network transmitter, but the direction cannot be determined. Where the network of transmitters is dense, then the mobile device may be detected by two or more of them and this allows the position of the mobile device to be determined by triangulation, although this my only be to within tens of metres depending on the network topology and local factors affecting radio wave transmission. Even where the transmitted density is lower adjacent cells will overlap to ensure continuous coverage, otherwise calls would drop out in the gap between cells. The handover region is an area where mobile devices can be located to low tens of metres degree of accuracy. The handover points can be used as virtual detection locations and the mobile device unique identifier and time of detection can be reported. By mapping the handover points to the road network, the travel time between handover points of individual handsets can be obtained to create a data set similar to that from ALPR or Bluetooth/Wi-Fi but without the need for any equipment deployed in the field.

More accurate location of mobile devices can be achieved if the device is equipped with a satellite GPS receiver and tracks and reports its own position via the mobile data network. The mobile device will determine its position using the service provided by a network of satellites orbiting the earth. The position can then be uploaded to the communications network provider or to a location-based service provider such as navigation. The frequency of uploading data may be slower than the frequency with which locations of the mobile device are recorded but the output of this process is a set of time and location coordinates. This is sometimes referred to as a 'breadcrumb trail'.

As well as time and location coordinates other data may be associated with each data point. Examples include the vector of the mobile device (i.e. speed and heading), temperature, rate of acceleration and, if the device is connected to a vehicle, parameters relating to on-board systems such as whether the wipers or lights are on, maximum braking force since the last reading and so forth. This additional data can provide added context for use in determining the presence of characteristics that indicate an incident has occurred or the device has encountered the consequences of an incident. Some emergency call (eCALL) services automatically trigger an alert at the service centre is the airbag is triggered. Subscribers to some satellite navigation services receive real-time warnings of queuing traffic ahead and advice on alternative routes if there is one with a shorter travel time, and shortest time is the selected criteria for route choice. However, the cause of the delay is not determined and whether or not that cause is an incident is not currently reported to the subscriber.

11.6.7   Social media and crowd-sourcing techniques

The techniques described so far do not involve any action by drivers or passengers but the reality is that some of them are also communicating using social media apps on smartphones across the 4G network. Dedicated apps such as WAZE provide a mechanism for drivers to report the presence of any problem they encounter via an interface designed to minimise driver distraction. The feedback includes some form of classification of the problem so that the broader context is known. WAZE is providing traffic advice in real time to its users as well as tracking them as probe vehicles. The user only has to select from a single menu to report an incident and the app automatically adds time, location and heading information before sending it to the WAZE servers. Google, who own and operate WAZE, have entered into data sharing agreements with a number of highway authorities so that the latter can access both the wide-area monitoring data and the incident reports. These sources can then be integrated into the authorities' traffic management system. On-board eCALL devices can also be triggered manually either by those in a vehicle involved in an incident or those in vehicles passing the scene.

  This approach has advantages in areas where there are sparse sensor information and little chance of detection by any other method. The drawbacks are that the number of vehicles using such apps in any section of road at any particular time may be very low. This means that any incident may not be reported or the time between the report and the actual incident may be long, possibly tens of minutes. Also, the reported location is where the user sent the report from. This may be some distance away from the actual incident. That said, some form of report is nearly always better than no report at all.

  A technique called 'sentiment analysis' has been developed to parse the messages from Twitter (tweets) and identify how Twitter users are reporting a particular experience. Sentiment Analysis is a field of natural language processing used to determine the attitude of the author of some document, be it a review or social media message. The goal of sentiment analysis is to find the senders view as being positive, neutral or negative from a given text. The rise of social media platforms has meant that there is a great interest in this field of analysis allowing for companies to leverage the data gathered to better market products, manage reputation and identify new opportunities. It has been used in public transport where the number of tweets with negative comments from passengers suddenly increases in response to a delay or problem. Public transport operators have been able to use this to identify problems on their network that have not been communicated through their own systems.

Using the Twitter application programming interface (API) all tweets mentioning given keywords or phrases can be retrieved for a particular period and in real-time. The queries that are processed by the API take the relevant keywords or phrases and filter those tweets down to the ones involving desired subject. Such keywords or phrases might include: 'incident', 'accident', 'roadworks', 'M75', etc. Once retrieved, the tweets contain information that allows us to filter down even further on data such as the user's location. However, most tweets are not geolocated so the keyword search needs to look for locations of interest, e.g. route number, junction number or name, as well as words related to traffic problems. A number of tweets within similar timeframes, problem descriptions and location description indicate the possible presence of an incident.

As with specialised apps such as WAZE, the main benefit of this approach is in areas where instrumentation is sparse. There are more Twitter users than users of traffic apps that support reporting but it is not guaranteed that they will report the events. Location is often poor as the phrase 'M4' or 'A7' can refer to several tens, even hundreds, of miles of a road. There is also the question of whether by using such information the traffic management authority is encouraging the use of twitter. In many jurisdictions, drivers would be committing an offence by entering a tweet into their smartphone while in control of a vehicle, even if it is stationary, although passengers would not.

Social media techniques are a useful adjunct to other methods and can direct a control room's attention to a problem that external sensors struggle to pick up. However, they are non-deterministic in the time to detect and the accuracy of location so are not suitable as a primary form of detection where safety is a major concern.

11.7   Comment on incident detection technology

The range of sensor technology available is increasing and their compactness, robustness and accuracy are improving while costs continue to fall. Low power requirements, improved battery technology combined with solar panel recharging and wireless communications are making the deployment of sensors easier and cheaper. As a consequence, the system provider is faced with complex choices and a detailed and well-thought-through-requirements specification is essential. Such a specification needs to recognise where trade-offs between cost, performance and long-term ownerships can be made. Well-designed pilot installations are also worthwhile, unless there is clear transferability of results from an existing user base. Any trials should be well-designed with a clear statistical basis and consider the system holistically including considerations of performance, ownership, scaling and business case. There is no single technology that suits all environments and it is for the system owner to invest the time and expertise to determine the solution that most closely achieves their objectives.

References

[1]   Weil, R., Wootton, J. and García-Ortiz, A. (1998). Traffic incident detection: Sensors and algorithms. Mathematical and Computer Modelling. Volume 27, Issues 9–11, pp. 257–291.

[2]   Williams, B. M. and Guin, A. (2007). Traffic management center use of incident detection algorithms: Findings of a nationwide survey. IEEE Transactions on Intelligent Transportation Systems. Volume 8, Issue 2, pp. 351–358.

[3]   Pursula, M. and Kosonen, I. (1989). Microprocessor and PC-based vehicle classification equipments using induction loops. Second International Conference on Road Traffic Monitoring. London, UK, 7–9 Feb. 1989, IET.

[4]   Dodsworth, J., Shepherd, S., and Liu, R. (2014). Real-time single detector vehicle classification. Transportation Research Procedia. pp. 942–951.

Chapter 12

Sensing of heavy precipitation—development of phased-array weather radar

Tomoo Ushio[1]

12.1 Introduction

With the advancement of society and global warming in recent years, tragic accidents caused by atmospheric phenomena such as tornadoes and heavy rainfall are on the rise. Development of phased-array Doppler meteorological radar that can observe meteorological phenomena such as cumulonimbus clouds that cause such local heavy rain and tornado in three dimensions with high resolution in time and space is progressing. In such a radar, the time required for observation is dramatically improved in comparison with the conventional method by using the electronic scanning method as compared to the conventional mechanical scanning method, and the world's best performance is realized. In this chapter, we outline the phased-array meteorological radar and its observation results and introduce future approaches to the use of local governments and others.

12.2 Background

In recent years, the damage caused by sudden heavy rain has been on the rise. For example, in a tragic flood accident that occurred on July 28, 2008 in Toga River, Kobe, a sudden, localized torrential rain occurs. In this accident, the 16 people were swept away due to the rapid rise in water level, and 5 died, including 2 primary school children and 1 nursery child. Alternatively, in Tsukuba City, Ibaraki Prefecture, an F3-scale tornado occurred on May 6, 2012, causing 30 injuries and 1 death. In the United States, hundreds of people are often killed due to tornadoes. In recent reports, there are an increasing number of hazardous phenomena such as localized heavy rainfall and tornadoes. For example, Figure 12.1 shows the time sequences in the number of annual occurrences of 50 mm or more per hour in Japan. It can be seen that although there is an increase or decrease depending on the year, it tends to increase as a whole. This is said to be the impact of rapid urbanization and global warming, but the true point is unknown.

[1]Graduate School of Engineering, Osaka University, Osaka, Japan

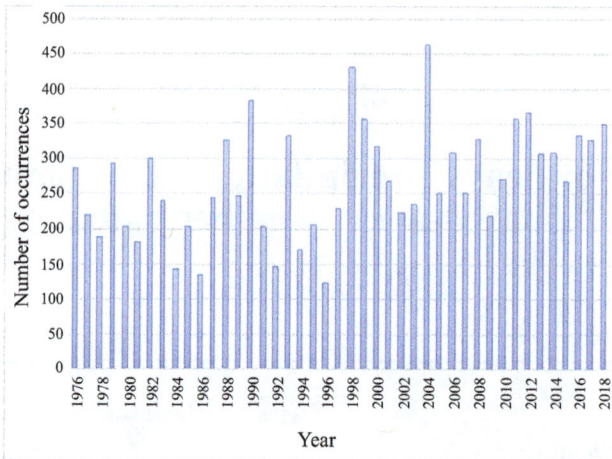

Figure 12.1   *Time series of the number of occurrences exceeding 50 mm/h (JMA website: https://www.data.jma.go.jp/cpdinfo/extreme/extreme_ p.html)*

The most effective means to measure such phenomena is remote-sensing technology using electromagnetic waves, which is well known as radar technology. The advantage of this method is that the structure of rainfall distributed over a wide area of dozens or hundreds of kilometers can be captured instantaneously by using an electromagnetic tool. For this reason, the Ministry of Land, Infrastructure, and Transport; the Japan Meteorological Agency; etc. have a large radar observation network to cover the whole of Japan, and we can also know the distribution of rainfall on the web, etc. in real time. And, using the rainfall distribution map captured by such a radar observation network, local governments, etc. are used for judgments such as evacuation instructions.

12.3   Problems

Radar (RADAR), which is a typical example of applied electromagnetics or remote-sensing technology, is an abbreviation of radio detection and ranging, in which electromagnetic waves emitted from a transmitting antenna are received after being scattered or reflected by a target. The principle is to obtain information on the distance to the targets and shape from the time difference between transmitting and receiving and the amplitude of the received signals.

The current radar based on such a principle uses a parabolic-type antenna and observes an area within a narrow beam width around 1 degree while gradually raising the elevation angle while rotating 360 degrees in the azimuth direction. However, with this method, it takes about 1–5 min for scanning near the ground alone, and 5–10 min or more for 3D observation. On the other hand, cumulonimbus clouds that cause the abovementioned localized torrential rain develop rapidly in about 10 min, and a tornado occurs and moves in only a few minutes. It is difficult to observe the process from generation to development and decay continuously one after another. This is one of the major factors that impede the elucidation of the formation mechanism of these atmospheric phenomena, the finding of predictive phenomena, prompt warning, and prediction.

12.4   Phased-array weather radar

Under these circumstances, Toshiba, National Institute of Communications and Technology (NICT), and Osaka University's Industry–Government–Academia Collaboration Group took the electronic/software scanning method instead of the mechanical antenna scanning and succeeded in developing an X-band phased-array Doppler weather radar that significantly reduces time for observations and enables to capture three-dimensional precipitation structure in detail in a very short time of 10–30 s (Figure 12.2) [1–3].

Phased-array radar does not point a parabolic antenna to the target as in the conventional radar, but by controlling the phases of the signals on the circuit quickly, the electromagnetic waves are radiated from many small antennas arranged in a plane. Since it is possible to form a beam at high speed, it is suitable for high-speed scanning as compared to mechanically driven.

Figure 12.3 shows the antenna part of the radar. Below the panel on this plane, 128 antennas are arranged, and 24 elements out of the 128 antenna elements are used to transmit a relatively wide transmit beam of around 10 degrees near the ground surface in the elevation direction. This transmission is repeated several times up to 90 degrees near the zenith direction. This is the phased-array beam transmission. Then, after receiving scattered waves scattered backward by the volume of the precipitation particles with the antenna of each of 128 elements, it makes the receiving

Figure 12.2   Photo of the phased-array weather radar in Osaka University

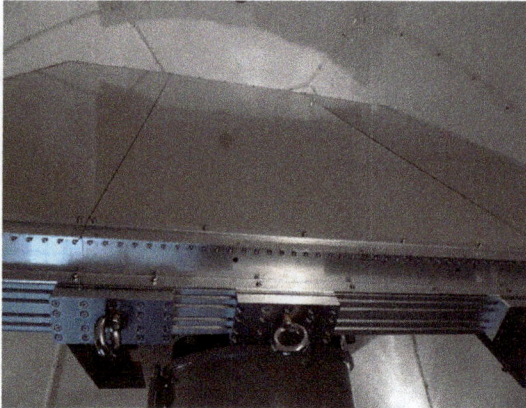

Figure 12.3   Antenna of the phased-array weather radar inside the radar dome

beam narrow after performing digital conversion and obtains about 1 degree receiving beam width. This is called digital beamforming (DBF) technology. The combination of this phased-array and DBF technology has made it possible to scan at high speed over 100 elevation angles without mechanically moving the antenna in the elevation direction. After that, a detailed three-dimensional precipitation distribution without gaps in the range from a radius of about 15–60 km and an altitude of 15 km is observed in just 10–30 s by mechanically rotating in the azimuth direction.

12.5   Observations

First phased-array weather radar at X band was installed on the roof of the E3 building at the Osaka University's Suita Campus in May 2012, and observation has been conducted continuously throughout the year until now. In the meantime, various types of precipitation events and cumulonimbus clouds were observed. Here, as an example of observation, heavy precipitation event observed shortly after the radar installation (Figure 12.4) is shown. This is an example observed on July 6, 2012. It was an event in which a linear precipitation system passed from the west to the east of the Osaka University's Suita Campus where the radar was installed. The right figure shows a cross section at an elevation angle of 4.35 degrees, and the left figure shows a vertical cross section in the direction west from the radar (center of the circle in the right figure). First, it can be seen that linear echoes are distributed from the center of the circle of the radar installation location to the west to the north. And, if you look at the left figure that is the vertical cross section, this precipitation system is well developed up to about 10 km in height, and the core of reddish precipitation with high-precipitation intensity is formed near the altitude of 5 km. And as time goes by, this precipitation core moves horizontally very fast and gradually falls down to the ground. Thus, in the present phased-array weather radar, such a falling precipitation can be seen as a three-dimensionally dense image. This is a

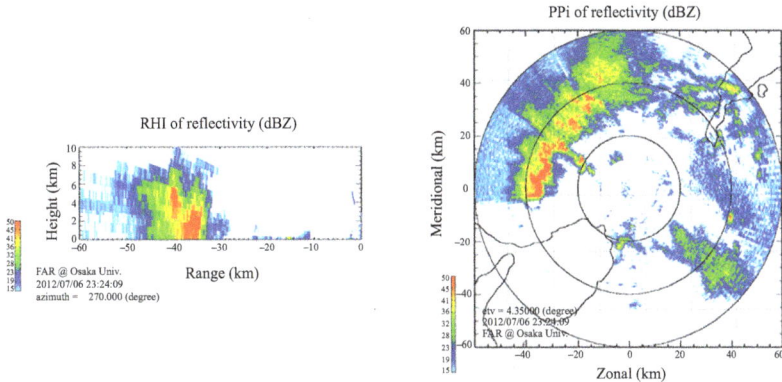

Figure 12.4   Observed image of the thunderstorm in July 6, 2012. The left panel shows the vertical cross section, and the right panel shows PPI at 4.35 degrees in elevation

Figure 12.5   Photo of the dual-polarimetric phased-array weather radar in Tokyo metropolitan area

great advantage of this phased-array radar, and the rapid scanning makes it possible to follow the generation, enhancement, and falling process of the precipitation core as a continuous image. In this way, the developed phased-array radar succeeded in capturing the behavior of cumulonimbus clouds that were generated, developed, and changed in a short time.

Furthermore, a dual-polarization phased-array weather radar was developed in 2018, which added a polarization observation function to this phased-array radar. Figure 12.5 shows a photograph of the dual-polarization phased-array weather radar at X band in Tokyo metropolitan area. With dual-polarized radar that combines measurement of vertical and horizontal polarization of electromagnetic wave, more accurate precipitation estimation and identification of precipitation particles are possible. Figure 12.6 shows the comparison of the radar reflectivity factor between a dual-polarization phased-array meteorological radar and a parabola-type radar installed several kilometers away. Thus, it is shown that although both radars have different transmitting and receiving systems, almost equivalent observation is performed. Thus, the phased-array meteorological radar has the ability to rapidly

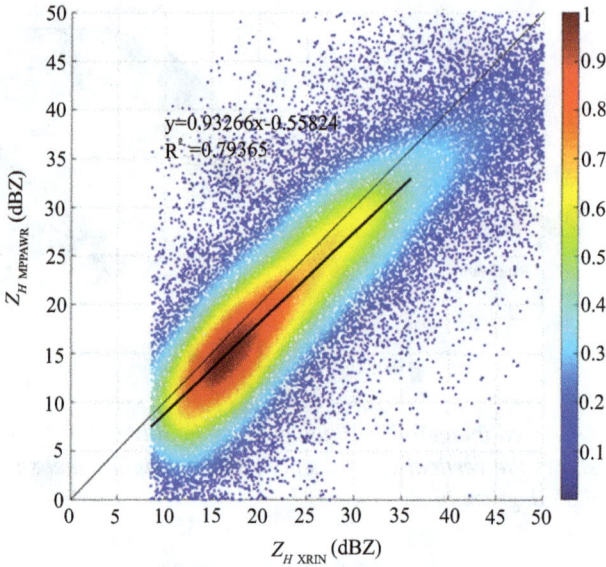

y=0.93266x-0.55824
R²=0.79365

*Figure 12.6   Scattergram of the radar reflectivity factor measured by dual
polarimetric phased-array weather radar and parabolic type radar
at X band colocated*

resolve the internal structure of the cumulonimbus cloud that causes massive damage
locally in a short time and provide effective data for short-time prediction. Taking
advantage of such phased-array meteorological radar, social experiments were
conducted to show the effectiveness of the phased-array weather radar. Figure 12.7
shows a schematic diagram of this social experiment system, and Figure 12.8 shows
the results of the experiment in a public park in Osaka. As shown in Figure 12.7, data
from the phased-array radar installed at Osaka University is transferred in real time
to a data center in Toshiba Kawasaki via the Internet. In this data center, not only
phased-array radar but also data from the Japan Meteorological Agency and the
Ministry of Land, Infrastructure, and Transport are simultaneously collected in real
time, and present rainfall rate, forecast of the rainfall rate, and warnings are issued,
which will be notified to various offices in local government in Osaka. In this case,
the warning issued by such a system is transmitted to the park management office at
16.09 on September 6, 2016, and in the park office, this warning information is
actually given to the visitors in the park through the speaker. The surveillance
camera in the park at 16.16 shows that the visitors are starting to evacuate, and at
16.21, lightning discharge is also detected around it, and rainfall exceeding 50 mm/h
was observed at the surface. Thus, in this system, it succeeded in issuing an alert
12 min before the actual heavy rain, which was impossible in the old system.
Actually, the old system issued the alert at 16.26, which is 17 min after the alert in the
phased-array system.

Real-time data flow architecture of Osaka
urban phased-array radar network

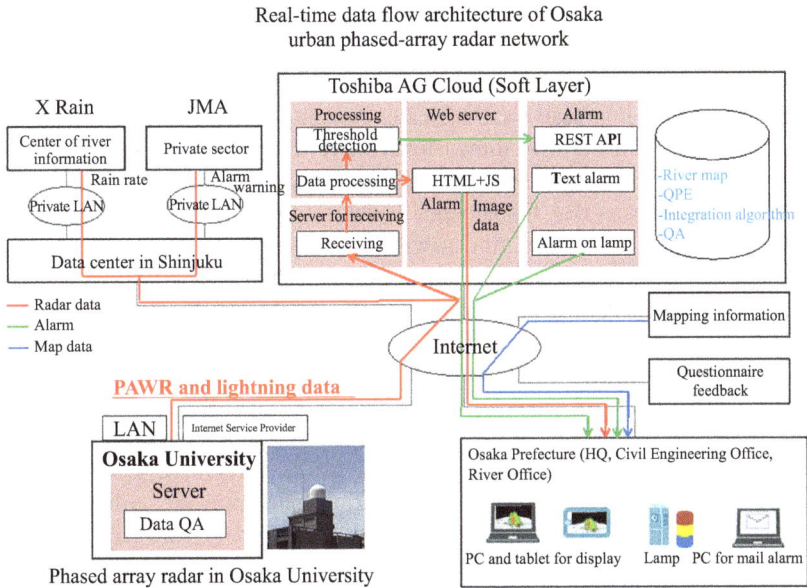

Figure 12.7   Real-time data flow architecture of Osaka urban phased-array radar network

Social experiment in public park

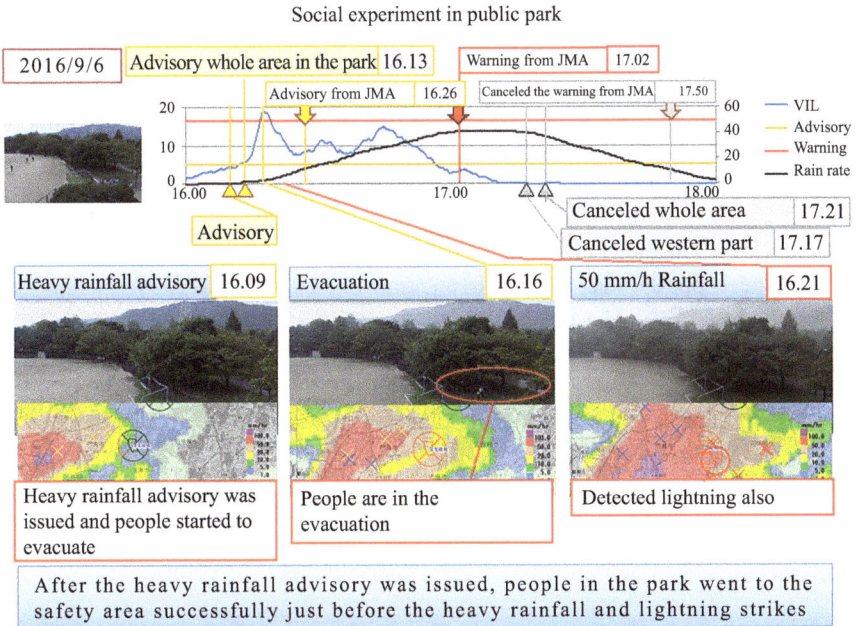

Figure 12.8   One example of the social experiments conducted at the public park in Osaka

12.6   Future

As we have seen earlier, the image or video shown by the data captured by this radar system is extremely impressive and indicates the potential level of this phased-array radar. First of all, detailed three-dimensional observation data obtained by this radar will clarify the mechanism of cumulonimbus cloud, which brings heavy rain in a short time. This means that fundamental scientific breakthroughs and discoveries will be made using this radar. And it will be applied to the improvement of the weather forecast, the detection of precursor phenomena of local and sudden weather disasters, and the short-time forecast (nowcast) information. It is thought that new utilization methods will be advanced in the transportation field such as train operation systems and safe operation of aircraft. Also, the same radar as this was installed in the NICT Kansai Advanced Research Center in Nishi-Kobe. As a result, the world's first phased-array radar network has been developed and tested. Such networked radar systems have significant advantages in several ways. Under such a network environment, more accurate precipitation estimation, etc. will be possible, and the robustness of the entire system will also be guaranteed. And such high-speed and high-resolution radar network is regarded as one high-precision and high-resolution super large-sized radar, and various applications are operated by each of the radar nodes in the network. This is a vision of the next generation of Japan's disaster prevention system and will be the most advanced and safest system in the world.

References

[1] Yoshikawa, E., T. Ushio, Z-I. Kawasaki, *et al.*, MMSE Beam Forming on Fast-Scanning Phased Array Weather Radar, *IEEE Trans. Geosci. Remote Sens.*, Vol. 51, Issue 5, pp. 3077–3088, 2013.

[2] Ushio, T., T. Wu, and S. Yoshida, Review of Recent Progress in Lightning and Thunderstorm Detection Techniques in Asia, *Atmos. Res.*, Vol. 154, Issue 1, pp. 89–102, doi:10.1016/j.atmosres. 2014.10.001, 2015.

[3] Mizutani, F., T. Ushio, E. Yoshikawa, *et al.*, Fast-Scanning Phased Array Weather Radar with Angular Imaging Technique, *IEEE. Trans. Geosci. Remote. Sens.*, Vol. 56, Issue 5, pp. 2664–2673, doi:10.1109/TGRS.2017.2780847, 2018.

Index

www.ingramcontent.com/pod-product-compliance
Lightning Source LLC
Chambersburg PA
CBHW060248230326
41458CB00094B/1542